E. COLI

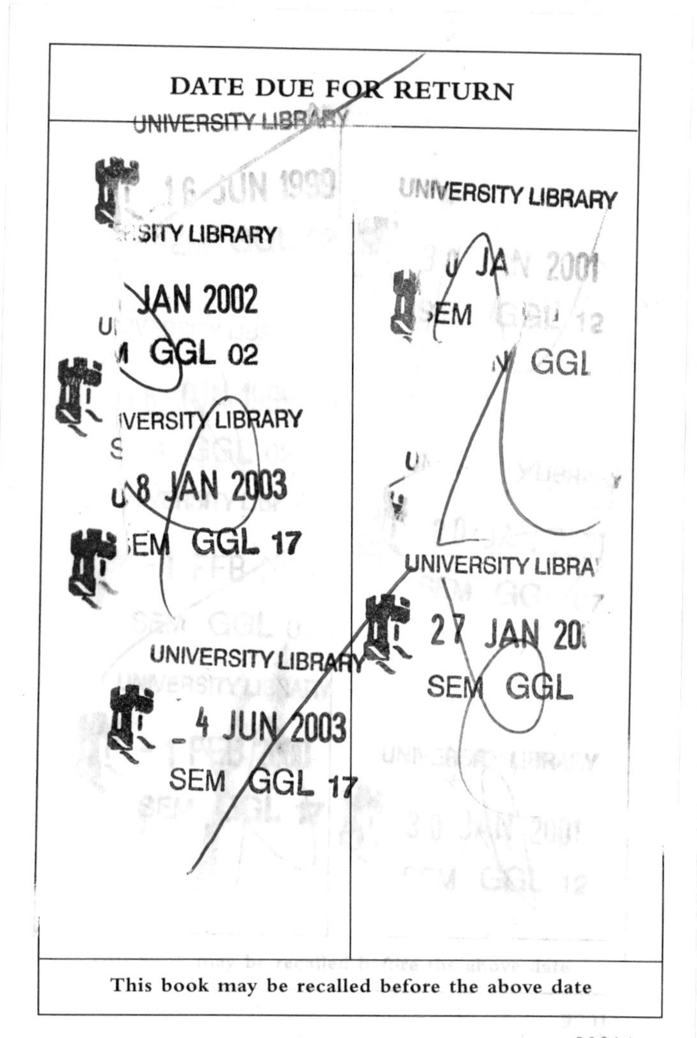

The Practical Food Microbiology Series has been devised to give practical and accurate information to industry about specific organisms of concern to public health. The titles in this series are:

Clostridium botulinum
A practical approach to the organism and its control in foods

E. coli
A practical approach to the organism and its control in foods

Listeria
A practical approach to the organism and its control in foods

Salmonella
A practical approach to the organism and its control in foods

To my lovely little terrors, Amelia and Oliver – a constant source of distraction (and sometimes pleasure!).

E. COLI

A practical approach to the organism and its control in foods

ERRATA SHEET IN BACK

Chris Bell
Consultant Food Microbiologist
UK

and

Alec Kyriakides
Company Microbiologist
Sainsbury's Supermarkets Ltd
London, UK

BLACKIE ACADEMIC & PROFESSIONAL
An Imprint of Chapman & Hall
London · Weinheim · New York · Tokyo · Melbourne · Madras

Published by Blackie Academic & Professional, an imprint of Thomson Science, 2–6 Boundary Row, London SE1 8HN, UK

Thomson Science, 2-6 Boundary Row, London SE1 8HN, UK

Thomson Science, 115 Fifth Avenue, New York, NY 10003, USA

Thomson Science, Suite 750, 400 Market Street, Philadelphia, PA 19106, USA

Thomson Science, Pappelallee 3, 69469 Weinheim, Germany

First edition 1998

Thomson Science is a division of International Thomson Publishing I(T)P®

Typeset in $10\frac{1}{2}/12\frac{1}{2}$pt ITC Garamond by Type Study, Scarborough, North Yorkshire

Printed in Great Britain by T. J. International Ltd, Padstow, Cornwall

ISBN 0 7514 0462 4

A catalogue record for this book is available from the British Library

Library of Congress Catalog Card Number: 98-72143

CONTENTS

FOREWORD

Escherichia coli was proposed as a human pathogen about a century ago but the importance of certain *E. coli* is greater now than at any previous time. The reason for this has been the emergence of Vero cytotoxin-producing *E. coli* (VTEC) as a major foodborne pathogen over the last 20 years. In particular, VTEC strains of serogroup O157 have caused very large outbreaks of food poisoning, often with significant morbidity and mortality. These have occurred in several different parts of the world, particularly in developed countries. This book provides a timely review of *E. coli*, particularly VTEC, in relation to foodborne disease and details the hazards in relation to specific foods and how these may be controlled by the food industry.

The book is divided into seven chapters complemented by over 50 tables and several figures. The introductory chapter covers the properties of *E. coli* in general, the diseases they cause and the reservoirs for these organisms. It also describes the diversity of *E. coli* in terms of virulence factors and how the different classes of *E. coli* are associated with particular serogroups and diseases. The second chapter describes major foodborne outbreaks caused by VTEC and, in one instance, by enteroinvasive *E. coli*. This section highlights the problems in food production that led to the outbreaks and the lessons to be learnt in each instance. Failures at different stages of food production are described, including cross-contamination of cooked products and *E. coli* surviving the manufacturing process or cooking. The hazards associated with sprouting vegetables and unpasteurized juice are also well documented. Chapter 3 deals with those factors affecting the growth and survival of *E. coli* and particularly *E. coli* O157. Several of the distinctive features of *E. coli* O157 are considered, for example its ability to survive under conditions of low pH, and it is clear that more work needs to be done in this area in relation to VTEC. The next chapter provides a comprehensive series of questions and answers for a range of food products linked to outbreaks caused by *E. coli* and particularly VTEC. This leads to an evaluation of the level of concern, full hazard analysis and how control measures should be applied. There is particular emphasis on raw fermented and dry-cured meat products, and foods that are produced and consumed without steps to eliminate any bacteria present. Chapter 5 deals with legislation and standards, including the complexities of European Union Directives as well as discussion of the 'zero-tolerance' approach to *E. coli* O157:H7 in raw ground beef adopted in the USA. Methods for detection are summarized in Chapter 6, including an extensive table covering information in this fast-moving area. In the final chapter the authors indicate some of the likely areas for study in future.

This book has a comprehensive, yet admirably concise, approach to the problems posed to the food industry by pathogenic *E. coli* and particularly by VTEC of serogroup O157. It highlights some of the many gaps in our knowledge at present but also provides a very helpful basis for understanding and controlling what is a major concern in relation to human health and foodborne disease.

Henry R. Smith
Laboratory of Enteric Pathogens
Central Public Health Laboratory
London

1

BACKGROUND

INTRODUCTION

Escherichia coli (*E. coli*) was discovered in 1885 by Dr Theodor Escherich (after whom the organism was ultimately named) during his work on bacteria in infant stools. Since its discovery, *E. coli* has become the workhorse of bacteriologists. This is largely because it is easy to grow, manipulate and characterize; consequently, the organism has been widely used in microbial genetics for, among other purposes, cloning the genetic material from many other organisms as part of procedures for learning more about the mechanisms for their control.

The importance of *E. coli* as a human pathogen has been recognized virtually since its discovery and the organism has been associated with diarrhoea (particularly in children), haemorrhagic colitis (HC), dysentery, bladder and kidney infections, surgical wound infection, septicaemia, haemolytic uraemic syndrome (HUS), pneumonia and meningitis; some of these conditions resulting in fatality. Generally, different strains of *E. coli* are associated with different clinical conditions.

The food industry also recognizes the importance of the organism and since the early 1900s it has been used as an indicator of faecal contamination in water and milk. The inclusion of *E. coli* in many food product specifications today also recognizes its value as an indicator of the hygienic status of food. Although widely accepted as a foodborne pathogen, it is only in more recent years that the food industry has refocused attention on *E. coli* as a cause of significant morbidity and mortality in outbreaks of foodborne illness; Vero cytotoxin-producing (Vero cytotoxigenic) *E. coli* (VTEC) being of particular and major concern.

Although there is still much to learn concerning the epidemiology of the organism, there are clear actions that can be taken by the food industry to minimize the incidence and level of the organism in foods, thus enhancing their safety.

This book aims to give the reader an overview of *E. coli*, particularly VTEC, but is primarily intended as an aid for those persons who want to understand the nature of the hazard it presents to food products and the means for controlling it.

TAXONOMY OF *E. COLI*

E. coli is by no means a new organism. In 1885 Theodor Escherich described some organisms he had isolated from infant stools. One of these he named *Bacterium coli commune*, also referred to as *Bacillus coli communis* in early texts (Lehmann, 1893).

For many years the generic term *Bacterium* was used to describe the broad group of Gram-negative, non-sporing rods occurring in the intestinal tract of man and animals, on plants and in the soil, and leading either a saprophytic, commensal or pathogenic existence.

By 1964, the genus *Escherichia* was defined in Wilson and Miles' *Topley and Wilson's Principles of Bacteriology and Immunity* as motile or non-motile organisms conforming to the definition of Enterobacterieæ (*sic*) as follows: Gram-negative, non-sporing rods; often motile, with peritrichate flagella. Easily cultivable on ordinary laboratory media. Aerobic and facultatively anaerobic. All species ferment glucose with the formation of acid or acid and gas, both aerobically and anaerobically. All reduce nitrates to nitrites. Oxidase negative; catalase positive. Typically intestinal parasites of man and animals, though some species may occur in other parts of the body, on plants and in the soil. Many species are pathogenic and give the biochemical characters shown in Table 1.1.

20 years later the genus *Escherichia* was clearly established in the family Enterobacteriaceae (Ørskov, 1984) and the main biotype characteristics of *E. coli* defined (Table 1.2). In 1987 the type genus *Escherichia* contained four species in addition to *E. coli* (Jones, 1988): *E. blattae* (isolated from the hind gut of the cockroach *Blatta orientalis*), *E. fergusonii*, *E. hermannii* and *E. vulneris*. All of these are biochemically distinct from *E. coli* (Farmer *et al.*, 1985).

Figure 1.1 shows the DNA relatedness of *E. coli* to some other genera of the Enterobacteriaceae, particularly some notable human pathogens. Based on DNA homology, *E. coli* and the four species of the genus *Shi-gella* should be considered as a single species (Jones, 1988). In common with

Table 1.1 Biochemical characteristics of *Escherichia* (Wilson and Miles, 1964)

Characteristic	Reaction
Mannitol fermentation	+, usually with gas
Lactose 37°C	Acid +, gas +
Lactose 44°C	Acid +, gas +
Adonitol	Seldom fermented
Inositol	Seldom fermented
Indole 37°C	Usually produced
Indole 44°C	Usually produced
Methyl red reaction	+
Voges–Proskauer reaction	−
Koser's citrate medium	−
Urea	No hydrolysis
Phenylalanine deamination	−
Kligler's H_2S (hydrogen sulphide) medium	No blackening
Møllers KCN (potassium cyanide) medium	No growth
Gluconate oxidation	−
Lysine decarboxylase	+
Glutamic acid decarboxylase	+
Gelatin liquefaction	−

+ = positive reaction; − = negative reaction.

work started over 70 years ago on the use of antigenic variation for strain differentiation in genera such as *Salmonella* and *Shigella*, the serological reactions of *Bacterium coli* were also studied. Such work was based on differences in the antigens on the surface of the bacterial cell, i.e. O (outer membrane) antigens, H (flagella) antigens and K (extracellular envelope (capsule)) antigens, where these are present. *Bacterium coli* was noted to be an antigenically heterogeneous species (Topley and Wilson, 1929a). In work reported in the early 1920s the evidence suggested that in some cases, such as in the haemolytic strains of the organism isolated from cases of acute urinary infection, antigenically homogenous groups seemed to be related to particular infective conditions.

Since the 1920s a great deal more work has been carried out on serogrouping within *E. coli* and in the mid 1940s a classification scheme was developed that allowed *E. coli* to be divided into more than 170 different serogroups based on the somatic (O) antigens (Kauffmann, 1947). In addition, over 50 flagella (H) antigens and approximately 100 capsular (K) antigens (previously divided into L, A and B antigens) are now also recognized and these are used to further subdivide *E. coli* into serotypes. Serogrouping and serotyping, together with other information such as

Table 1.2 Some key biotype characteristics of *E. coli* (adapted from Ørskov, 1984)

Characteristic	Reaction
Gram	Negative
Cell morphology	Non-sporing straight rod, 1.1–1.5 × 2.0–6.0 μm
Motility	+ by peritrichous flagellae or non motile
Aerobic growth	+
Anaerobic growth	+
Optimum growth temperature	37°C
Catalase	+
Oxidase	−
D-mannitol fermentation	⩾ 90% +
Lactose 37°C	⩾ 90% +
Lactose 44°C	⩾ 90% +
D-adonitol	⩾ 90% −
D-glucose	Acid produced
Indole 37°C	⩾ 90% +
Indole 44°C	⩾ 90% +
Methyl red reaction	⩾ 90% +
Voges–Proskauer reaction	⩾ 90% −
Growth in Simmons' citrate	⩾ 90% −
Urease, Christensen's	⩾ 90% −
Phenylalanine deamination	⩾ 90% −
Lysine decarboxylase	76–89% strains positive
H_2S in TSI (triple sugar iron) medium	⩾ 90% −
Growth in KCN medium	⩾ 90% −
Gelatin liquefaction (at 22°C)	⩾ 90% −

+ = positive reaction; − = negative reaction.

biotype, phage type and enterotoxin production, can now distinguish those strains able to cause infectious disease in man and animals (Linton and Hinton, 1988), and some correlation has been established between the *E. coli* serogroup and virulence, thus confirming the observations of the pioneers in this field.

Many types of disease are caused by *E. coli* depending on the virulence factors expressed. Some virulence factors identified in pathogenic *E. coli* serotypes include possession of adhesins or colonization factors, ability to invade epithelial cells of the small intestine, haemolysin production and toxin production (heat stable, ST; heat labile, LT; Vero cytotoxin 1, VT1; Vero cytotoxin 2, VT2). VT1 and VT2 are also referred to as shiga-like toxins (SLT) 1 and 2. There are currently six virulence types of *E. coli* recognized.

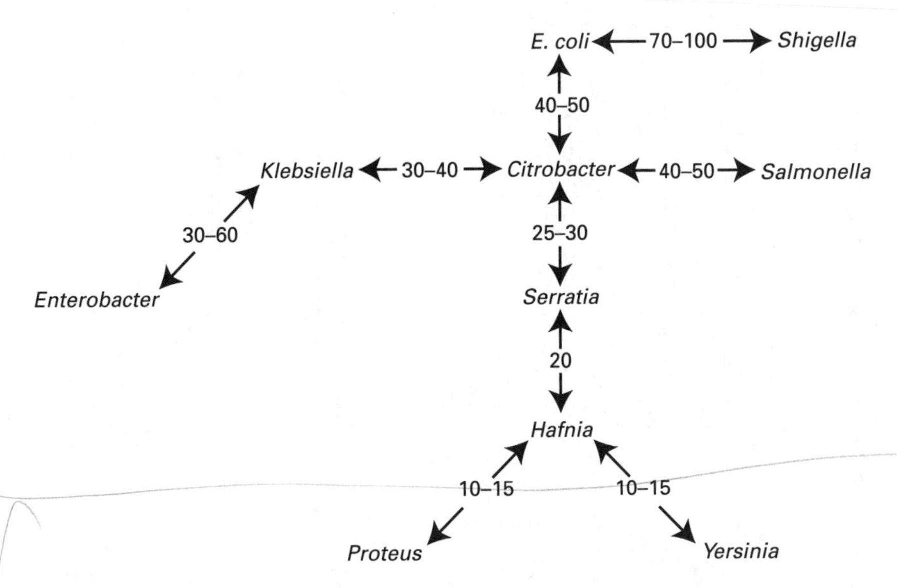

Figure 1.1 DNA relatedness of *E. coli* and some other genera of the Enterobacteriaceae (numbers represent approximate percentage relatedness). Adapted from Brenner, 1984.

Table 1.3 indicates these together with some *E. coli* serogroups and the types of diseases associated with these organisms.

ILLNESSES CAUSED BY *E. COLI*

Since Escherich first described *Bacterium coli commune*, the organism later to bear his name as *Escherichia coli*, its association with human disease has been the subject of enormous amounts of work, which continues today.

Concerning this organism Lehmann (1893) noted 'animals, especially guinea-pigs, die after the injection of large doses (subcutaneous or interveinous) with hæmorrhagic and necrotic processes, especially in the intestinal canal and diarrhœa; swelling of the spleen is not observed. Recently it has been repeatedly recognised as an exciter of dangerous and even fatal infections in man; in other cases it is a secondary intruder'(*sic*).

In 1921 Muir and Ritchie described the pathogenic properties of *B. coli* as follows,'in man, the b. coli has been found as the only organism present in various suppurative conditions, especially in connection with the intestine (e.g. appendicitis) and about the urinary tract. . . The b. coli is also

Table 1.3 Serogroups and disease associations of six virulence types of *E. coli* (Rowe, 1983; Salyers and Whitt, 1994; Beutin *et al.*, 1997; Sussman, 1997; Willshaw *et al.*, 1997)

Virulence type	Serogroup examples*	Disease association	Summary of *E. coli*/host interaction
Enteropathogenic (EPEC)	O18ab, O18ac, O26, O44, O55, O86, O111, O114, O119, O125, O126, O127, O128, O142, O158	Enteritis in infants Traveller's diarrhoea	EPEC attach to intestinal mucosal cells causing cell structure alterations (attaching and effacing) EPEC cells invade the mucosal cells
Enterotoxigenic (ETEC)	O6, O8, O15, O25, O27, O63, O78, O115, O148, O153, O159	Diarrhoea, vomiting and fever Traveller's diarrhoea	ETEC adhere to the small intestinal mucosa and produce toxins that act on the mucosal cells
Vero cytotoxigenic (VTEC) (includes Enterohaemorrhagic, EHEC)	O2, O4, O5, O6, O8, O15, O18, O22, O23, O26, O55, O75, O91, O103, O104, O105, O111, O113, O114, O117, O118, O121, O128ab, O145, O153, O163, O157, O168	Shigella-like dysentery (stools contain blood and mucus) Haemolytic uraemic syndrome	EHEC attach to and efface mucosal cells and produce toxin
Enteroinvasive (EIEC)	O28ac, O29, O112ac, O121, O124, O135, O136, O143, O144, O152, O164, O167, O173	Shigella-like dysentery	EIEC invade cells in the colon and spread laterally, cell to cell
Enteroaggregative (EAggEC)	Not yet established	Persistent diarrhoea in children	EAggEC bind in clumps (aggregates) to cells of the small intestine and produce toxins
Diffusely adherent (DAEC)	Not yet established	Childhood diarrhoea	Fimbrial and non-fimbrial adhesins identified

* Only certain strains within a serogroup may be associated with human illness and further characterization, e.g. serotype or phage type, is necessary to establish a clear identification of the causative agent.

apparently the cause of some cases of summer diarrhœa (cholera nostras), of some cases of infantile diarrhœa, and of some food poisonings'(*sic*).

In Topley and Wilson's first edition of *The Principles of Bacteriology and Immunity* (1929a) the pathogenicity of *Bacterium coli* was summarized: '*Bact. coli* is a normal inhabitant of the intestine of man and other animals. In certain circumstances it acquires pathogenicity, and may cause local or general infection. It is a frequent cause of acute and chronic infection of the urinary tract, and may give rise to an acute or chronic cholecystitis.'

Today it is known that *E. coli* is commonly a harmless member of the normal commensal microflora of the distal (end or terminal) part of the intestinal tract of humans and other warm-blooded animals, comprising less than 1% of this population in numbers ranging up to 10^2 per gram of faeces in humans (Smith, 1961). *E. coli* is acquired by infants within a very few days of birth. The organism is acquired predominantly from the mother by the faecal–oral route but also via environmental surroundings. Although most strains of *E. coli* are not pathogenic, the species does contain strains that can cause a number of different types of illness, some fatal, and some of these strains are known to be foodborne. Table 1.4 summarizes the clinical characteristics of some of these important types of *E. coli*.

Oral challenge experiments have been carried out with some *E. coli* types and the results of such studies suggest that levels of 10^5 to 10^{10} EPEC organisms are required to produce diarrhoea, 10^8 to 10^{10} ETEC organisms are necessary for infection and diarrhoea and 10^8 cells of EIEC are required to produce diarrhoeal symptoms in normal adults. However, this may vary depending on the acidity in the stomach as severe diarrhoea developed following the ingestion of only 10^6 cells of EIEC 5 min after volunteers were administered 2 g of sodium bicarbonate (Doyle and Padhye, 1989; Sussman, 1997). Epidemiological evidence and studies of foods associated with human illness suggests that the infective dose of Vero cytotoxigenic *E. coli* can be very low, e.g. <100 cells of the organism (Advisory Committee on the Microbiological Safety of Food, 1995; Bolton et al, 1996).

The haemolytic uraemic syndrome caused by VTEC and characterized by acute renal failure, haemolytic anaemia and thrombocytopaenia usually occurs in young children (under 5 years of age). It is the major cause of acute renal failure in children in the UK and several other countries. Up to 10% of patients infected with VTEC O157 develop HUS and some adult patients develop thrombotic thrombocytopaenic purpura (TTP). Infection

Table 1.4 Characteristics of *E. coli*-related illness (adapted from Doyle and Padhye, 1989)

Pathogenic type of *E. coli*	Time to onset of illness	Duration of illness	Range of symptoms
EPEC	17–72 h, average 36 h	6 h to 3 days average 24 h	Severe diarrhoea in infants, which may persist for more than 14 days Also, fever, vomiting and abdominal pain In adults, severe watery diarrhoea with prominent amounts of mucus (main symptom) without blood, and with nausea, vomiting, abdominal cramps, headache, fever and chills
ETEC	8–44 h, average 26 h	3–19 days	Watery diarrhoea, low-grade fever, abdominal cramps, malaise, nausea When severe, causes cholera-like extreme diarrhoea with rice-water-like stools, leading to dehydration
VTEC	3–9 days, average 4 days	2–9 days, average 4 days	Haemorrhagic colitis (HC): sudden onset of severe crampy abdominal pain, grossly bloody diarrhoea, vomiting, no fever Haemolytic uraemic syndrome (HUS): prodrome of bloody diarrhoea, acute renal failure in children, thrombocytopaenia, acute nephropathy, seizures, coma, death Thrombotic thrombocytopaenic purpura (TTP): similar to HUS but also fever, central nervous system disorders, abdominal pain, gastro-intestinal haemorrhage, blood clots in the brain, death
EIEC	8–24 h, average 11 h	Days to weeks	Profuse diarrhoea or dysentery, chills, fever, headache, muscular pain, abdominal cramps

with VTEC O157 can result in death and rates of fatality vary considerably but can be high, e.g. over 9%, particularly where institutional outbreaks occur (Stewart *et al.*, 1997). Generally, about 5% of cases of haemorrhagic colitis caused by VTEC progress to HUS, in which case the fatality rate is approximately 10% (Anon., 1995a). The diarrhoeal phase of the infection caused by *E. coli* O157 is usually self-limiting and there are no specific treatments of the conditions caused by the organism. Each symptom is treated as it occurs in the individual (Advisory Committee on the Microbiological Safety of Food, 1995; Anon., 1995b). The usefulness of antibiotics in controlling the course of the illness is not clear and is likely to be complicated by the growing evidence of an increasing prevalence of antimicrobial resistant strains of VTEC O157 (Willshaw *et al.*, 1997). Such information is of concern to the food industry, which is already having to deal with the implications of multiple antibiotic resistant strains of *Salmonella typhimurium* DT104.

World-wide, diarrhoeal and other diseases caused by *E. coli* are very important, particularly in children. Because of the potential for Vero cytotoxigenic *E. coli*, particularly serogroup O157, to cause HUS and HC in very young children and, to a lesser extent, the elderly, and the increasing number of different food vehicles such as beefburgers, yoghurt, apple cider implicated in infections with *E. coli* O157, a great deal of attention and work has been focused on this organism over the past 10 years. Table 1.5 summarizes information concerning some food-associated outbreaks caused by different types of *E. coli*. Such information underlines the international importance of *E. coli* infections and the wide-ranging role of food and water as sources of the organism. The large number of secondary cases associated with many outbreaks also indicates the importance of person-to-person spread by the faecal–oral route. Outbreaks of infection in hospitals, child-care centres and nursing homes have been attributed to this route (Advisory Committee on the Microbiological Safety of Food, 1995). Young children have been shown to carry VTEC O157:H7 longer than older children or adults after the symptoms have resolved; carriage rates of over 3 weeks have been observed in children under 5 years of age (Griffin, 1995).

Information concerning the incidence of VTEC in populations, and geographical and seasonal distributions is incomplete; however, current data suggest that infections are more common in the USA, Canada and the UK (possibly reflecting the structured national epidemiological and micro-biological surveillance systems in place in these countries for identifying and reporting disease outbreaks, which may be unrecorded in other countries).

Table 1.5 Examples of food-associated outbreaks of disease caused by *E. coli*

E. coli type	Year	Country	Suspected food vehicle	Cases (deaths)
EPEC O86:B7:H34[a]	1961	Romania	Coffee substitute	10 (0)
EPEC O111:B4[a]	1967	Washington DC, USA	Water	170 (0)
EPEC O111:B4[b]	1987	Finland	?	787 (0)
ETEC O27:H20[a,c]	1983	USA, also Denmark, Netherlands, Sweden	French Brie cheese	169 (0) (in USA)
ETEC O6:K15:H16[a]	1975	Oregon, USA	Water	> 2000
ETEC O6:H16 (ST^+, LT^+) and O27:H20 (ST^+)[d]	1983	UK	Curried turkey mayonnaise	27 (0)
EIEC O124[a]	1947	UK	Canned salmon	47
EIEC O124:B17[e]	1971	USA	French Brie and Camembert cheeses	387
EIEC non-typeable[a]	1981	USA (cruise ship)	Cold buffet, potato salad	?
VTEC O157:H7[f]	1982	Oregon and Michigan, USA	Hamburger patties in sandwiches	> 47 (0)
VTEC O157:H7[g]	1985	UK	Handling vegetables, particularly potatoes	> 24 (1)
VTEC O157[h]	1985	Ontario, Canada	Undercooked beef patties	73 (17)
VTEC O157:H7 phage type 49[i]	1991	UK	Yoghurt	16 (0)
VTEC O157:H7[j]	1993	USA	Hamburgers	732 (4)
VTEC O111:NM[k]	1995	South Australia	Uncooked, semi-dry fermented sausage (Mettwurst)	> 23 (1)
VTEC O157[l]	1996	Sakai City, Japan	White radish sprouts	6309 (3)
VTEC O157 phage type 2[m]	1996	Scotland, UK	Meat products	490 (20)

[a] Doyle and Padhye, 1989
[b] Viljanen *et al.*, 1990
[c] MacDonald *et al.*, 1985
[d] Riordan *et al.*, 1985
[e] Marier *et al.*, 1973
[f] Riley *et al.*, 1983
[g] Morgan *et al.*, 1988
[h] Chapman, 1995
[i] Morgan *et al.*, 1993
[j] Advisory Committee on the Microbiological Safety of Food, 1995
[k] Cameron *et al.*, 1995a, b
[l] Fukushima *et al.*, 1997
[m] Pennington, 1997; Reid, 1997

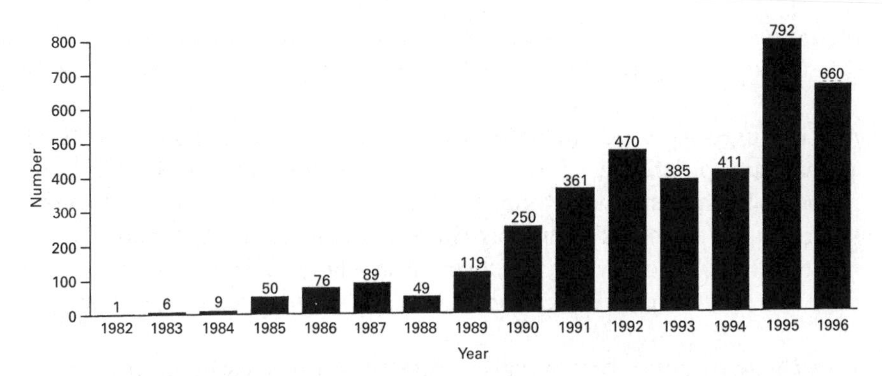

Figure 1.2 Laboratory-confirmed cases of VTEC O157 infections in England and Wales 1982 to 1996 (Anon., 1997a).

Also, infections clearly occur in a seasonal pattern: in late spring to late summer (UK) or in mid-summer to mid-autumn (USA) (Chapman, 1995).

From the early 1980s to 1996 the number of laboratory-confirmed cases of VTEC O157 reported in the UK has shown a steady increase (Figure 1.2). The cumulative total of laboratory reports of Vero cytotoxin-producing isolates of *E. coli* O157 for 1997 was 1060 and for the first 6 weeks of 1998,78 (34 for the same period in 1997; Anon., 1998a, b). Evidence from other countries such as the USA and Canada also suggests that infections with the organism have been increasing steadily.

Since the food association with infection by VTECs has been recognized and the severity of illness caused by infection with VTECs, particularly *E. coli* O157:H7, in vulnerable groups of the population has become increasingly understood, the food industry has focused considerable attention on controls to minimize contamination of food by the organism.

SOURCES OF *E. COLI*

The primary habitat of *E. coli* is the intestinal tract of man and other warm-blooded animals, and *E. coli* infections are transmitted via three main routes: directly from animals, including farm animals and domestic pets, person-to-person spread and contaminated foods.

Many serogroups of *E. coli* are found as normal and harmless inhabitants of the mammalian gut but some contain serotypes that are pathogenic to man and animals; these are not regarded as part of the normal flora of the human intestine. Domestic pets such as dogs and cats carry *E. coli*, including serogroups containing types pathogenic to humans, e.g. O55, O111 and O128 (Bettelheim, 1997).

Animals used for food, including cattle, pigs, sheep, the young of these animals and poultry, all carry *E. coli* as commensal flora, often different from the 'normal' strains in humans. They may also be infected by specific strains, again often different from those infecting humans. Strains pathogenic to humans carried in the 'normal' gut flora of food animals clearly pose a potential risk of infection to humans via a number of routes:

- faecal-oral route from animals to humans during rearing processes
- faecal contamination of food crops when untreated or poorly treated manure is used for fertilizer
- faecal contamination of carcasses via poor hygienic practices during slaughter and evisceration processes
- consumption of faecally contaminated raw milk, *E. coli* mastitic milk or products made from such milk.

Table 1.6 indicates the occurrence of *E. coli* O157 in some animals and the meat from animals reported in different surveys. Although the overall reported incidence of *E. coli* O157:H7 in bovine meats seems low (<4%), it was noted by Tilden *et al.* (1996) that in one outbreak counts of 10^3 *E. coli* O157:H7 per gram of meat were found on one carcass. Such levels distributed in comminuted meat products, e.g. burgers or sausages, can lead to low-level contamination of large numbers of individual products. Thus, cattle are now widely accepted as a major reservoir of VTEC O157:H7, currently the most important of the foodborne Vero cytotoxigenic *E. coli*.

In addition to the foods implicated in the outbreaks of human illness shown in Table 1.5, a wide variety of other foods also implicated in outbreaks clearly demonstrate the potential for many food types to become contaminated with disease-causing *E. coli* serotypes. In particular, serious illness due to VTEC O157 has been linked to consuming cantaloupe melon (Del Rosario and Beuchat, 1995), salad dressing containing mayonnaise made in-store (Zhao and Doyle, 1994), cooked ham (Gammie *et al.*, 1996) and unpasteurized apple cider (Besser *et al.*, 1993; Mshar *et al.*, 1997).

E. coli is commonly found in external environments (soil and water) that have been affected by human and animal activity, and the presence of *E. coli* in water systems has for many decades been used as an indicator of recent faecal contamination of water (Topley and Wilson, 1929b; Report 71, 1994). In a recent survey carried out in the UK of faecal samples from wild birds (mainly gulls) between 0.9 and 2.9% of the bacterial isolates

Table 1.6 Surveys of the occurrence of *E. coli* O157 and non-O157 VTEC

Subject	Country	Occurrence (%)	Reference
Retail fresh meats and poultry	USA	6/164 (3.7) ground beef 4/264 (1.5) pork 4/263 (1.5) poultry 4/205 (2.0) lamb All isolates *E. coli* O157:H7	Doyle and Schoeni, 1987
Retail chickens and sausages	UK	46/184 (25) pork sausages 0/71 (0) chickens All non-O157 VTEC	Smith *et al.*, 1991
Cattle	UK	84/2103 (4) bovine rectal swabs 7/23 (30) carcasses from rectal swab-positive cattle 2/25 (8) carcasses from rectal swab-negative cattle All isolates *E. coli* O157	Chapman *et al.*, 1993
Dairy herds	USA	6/399 (1.5) faecal specimens from calves 24 h old to weaning age 13/263 (4.9) calves weaned to 4 months old Positive for *E. coli* O157:H7	Zhao *et al.*, 1995
Retail raw meats	Netherlands	*Survey 1* 2/770 (0.3) minced mixed beef and pork 0/1000 (0) raw minced beef 0/260 (0) minced pork 0/300 (0) poultry products Isolates confirmed as VTECs *Survey 2* 6/201 (3.0) minced beef 15/110 (13.6) chicken products 1/49 (2.0) minced pork All isolates *E. coli* O157 but VT negative	Heuvelink *et al.*, 1996
Raw meat products	UK	3/89 (3.4) frozen beefburgers 1/50 (2.0) fresh minced beef 0/50 (0) fresh sausages All isolates *E. coli* O157	Bolton *et al.*, 1996

obtained were VTEC O157 (Wallace *et al.*, 1997), demonstrating the potential for seabirds that commonly forage on farm land to cycle serious human pathogens through the primary agricultural environment. In the UK, microbiological standards including *B. coli* were suggested for application in assessments of the bacterial pollution of milk (Savage, 1912) to be used indirectly as a means for improving the general supply. Today *E. coli* is still widely used as an indicator of faecal contamination of food and water, but in the food industry specific tests for the detection of VTEC, particularly O157:H7 in relevant foods, have become part of the due diligence procedures in place to monitor the effectiveness of food safety control systems.

The widespread nature of *E. coli* makes it inevitable that most raw food materials will be contaminated by the organism, albeit often at a low level. Raw foods, including meats, milk, vegetables and salad vegetables, will occasionally be contaminated with *E. coli* types capable of causing human infection, including VTEC O157:H7 which is being found in an increasingly wide range of animals.

2

OUTBREAKS: CAUSES AND LESSONS TO BE LEARNT

INTRODUCTION

Numerous food poisoning outbreaks and incidents that have been attributable to *E. coli* have been recorded during the past century. Table 1.5 summarizes information concerning some of these outbreaks and clearly highlights the widespread occurrence of the hazard and world-wide distribution of outbreaks. It also provides evidence of the diverse nature of foods implicated in such outbreaks.

It is interesting to note that until fairly recent times in the developed world, foodborne outbreaks of *E. coli* infection have not been accompanied by high levels of mortality. Although capable of causing severe infections, it was not until the outbreaks caused by Vero cytotoxin-producing *E. coli* that fatalities began to accumulate, with the concomitant public concern. The widespread nature of this hazard has resulted in most food groups being implicated in outbreaks, including dairy, meat and vegetable products, although a high number of additional outbreaks have been associated with the consumption of contaminated water. Analysis of outbreaks caused by all groups of *E. coli* identifies faecal contamination of foods or water as being the primary cause, due to either inadequate sanitation or poor standards of food processing or personal hygiene. However, when focusing on Vero cytotoxin-producing *E. coli*, and *E. coli* O157 in particular, foods that meet the following criteria are far more likely to be implicated in outbreaks than others:

- raw ingredients of bovine origin
- raw ingredients subject to direct/indirect contamination from bovine sources, i.e. manure or irrigation water
- raw ingredients not subsequently subjected to a bacterial destruction process

- products subject to post-process contamination
- processes that may allow the growth of VTEC, if present.

Outbreaks of food poisoning are a reminder that it is often the simple control measures in the food processing, retailing and catering sectors that, when inadequately controlled, can be the difference between safe and unsafe food. The following section reviews some of the outbreaks that have occurred in recent years to assess what may have gone wrong and what measures could be taken to prevent such incidents occurring again. Complete information relating to the outbreaks has not always been recorded because food poisoning is often investigated from a clinical perspective with details of the food characteristics, processing and environmental conditions not being documented. The information and comments in Tables 2.1 to 2.7 represent a combination of the known information together with the most likely reasons why these incidents occurred and ways of controlling them. Analysis of outbreaks in this way is a useful means of identifying potential deficiencies in production systems that may compromise the safety of products in relation to this hazard. It is hoped that by learning from the mistakes of others such information can be used to improve the safety of subsequent processes and products.

COOKED MEAT PRODUCTS: SCOTLAND

The largest number of fatalities experienced in a single outbreak of Vero cytotoxin-producing *E. coli* infection was recorded in Lanarkshire, Scotland in 1996–97. The outbreak was reported to have resulted in over 490 cases, of which 265 were confirmed as being caused by the outbreak strain of *E. coli* O157 (Reid, 1997). 20 people who suffered illness died, all of them being elderly patients, primarily from a local nursing home. Of the 107 affected individuals who were admitted to one hospital, more than 50% were over 60 years of age although patients included those from all age ranges (0–4 to 90+) (Stewart *et al.*, 1997). It was also reported that all deaths which occurred were elderly patients and were associated with haemolytic uraemic syndrome, renal failure or other complications. The strain causing the illness was reported to be *E. coli* O157 phage type 2 (PT2), VT1 negative, VT2 positive (Ahmed, 1997). Initial investigations indicated that most of the affected individuals had consumed cold cooked meat products (Table 2.1) from a butcher in Wishaw, Lanarkshire or had consumed a steak pie at a lunch supplied by the same butcher (Ahmed, 1997). *E. coli* O157 PT2 was isolated from food and environmental samples taken from the butcher's shop and the strains were considered to be indistinguishable from those isolated from infected human cases

Table 2.1 Outbreak overview: cooked meat

Product type:	Cooked meats or cooked meat products
Year:	1996–97
Country:	Scotland
Levels:	Present in finished products but levels not reported

Possible reasons

(i) Cross-contamination from raw meat to cooked meat products
(ii) Inadequate reheating of products to be eaten hot

Control options*

(i) Adequate cooking procedures at all stages
(ii) Training of staff in safe food hygiene practices
(iii) Effective segregation of raw and cooked meat products during manufacturing and retail handling
(iv) Operation of effective cleaning procedures

* Suggested controls are for guidance only and may not be appropriate for individual circumstances. It is recommended that proper hazard analysis is carried out for every process and product to identify where controls must be implemented to minimize the hazard from *E. coli*.

(Ahmed, 1997). The outbreak provoked widespread media attention and associated public concern, resulting in an independent inquiry which published wide ranging recommendations relating to raw and cooked meat hygiene standards, particularly in retail shops (Pennington, 1997).

The reasons for this outbreak have not been fully reported. However, the detection of *E. coli* O157 in food and environmental samples from the butcher's shop, together with the fact that cooked meats were implicated in the outbreak, indicate that either inadequate cooking occurred or products may have been contaminated with the organism due to post-process contamination. The butcher's shop implicated in this outbreak was believed to be a medium-scale manufacturer and supplier of raw and cooked meat products to the local community. The equipment and premises were believed to be of a high standard and the butcher was reported to have been awarded a 'Butcher of the Year' prize, although the extent to which this encompassed hygienic practices is not clear. The standards of food hygiene, staff training and process controls are not known but inevitably there is likely to have been a failure in one of these critical areas to allow such contamination of cooked meat products to occur.

Clearly, for this outbreak to have occurred raw materials, products, equipment or people contaminated with *E. coli* O157 must have provided source(s) of the organism. The most likely source of the organism is raw meat, particularly beef. Cattle have been identified as a source of *E. coli* and *E. coli* O157 in particular (Hancock *et al.*, 1997a). Contamination of animals or carcasses by faeces during either transportation, slaughter or processing has been the subject of much study. Whether arising from faecal contamination from the hide, during evisceration or due to discharge of gut contents; if the organism is present in cattle it is inevitable that some contamination will occur (Heuvelink *et al.*, 1996).

It is not possible to prevent the contamination of raw agricultural products with pathogens, particularly enteric pathogens, and from time to time their presence on these products must be anticipated. However, by exercising good practices in animal husbandry and high standards of slaughter hygiene it is possible to limit the frequency and level of contamination of carcasses with pathogens. Most local, small butcher shops are unlikely to be able to influence standards of slaughterhouse hygiene therefore any processor of raw meat, including butcher shops, must recognize that raw meats and meat products will carry enteric pathogens such as *E. coli* O157. Relevant controls must therefore be in place and operated at the processing premises to reduce the hazard during the process and prevent recontamination of cooked products.

Controlling cross-contamination in an area handling both raw and cooked meat products is a difficult task. It is not known whether staff in the implicated butcher's shop were dedicated to one task such as butchery or whether they were multi-functional, handling both raw and cooked products. Naturally, with dedicated staff it should be far simpler to prevent cross-contamination from raw to cooked products providing staff movements, equipment locations and work areas are carefully planned. Common entry points and passageways for staff handling raw meats and those handling cooked products should be avoided and areas for storing their work clothing should also be separated to prevent cross-contamination.

Raw meat should be prepared on dedicated surfaces, using dedicated utensils such as knives, slicing equipment and storage bowls, and ideally in a separate room from cooked meat handling. Areas often forgotten in these types of operation are storage areas such as refrigerators, where raw and cooked meat products are often held in close proximity. Ideally, separate refrigerators should be used for storing raw and cooked foods. However, in the absence of complete separation, designated areas should be identified for use for only raw or only cooked products. All products should be fully wrapped and basic food hygiene rules must be applied, e.g. raw foods should always be stored below cooked foods.

The fact that such a large outbreak occurred in a few localized areas may implicate contaminated bulk product. Assuming that the cooking process was effective, the most likely cause of contamination would have been post-cooking contamination either from equipment, including preparation surfaces such as tables, or personnel. Indeed, in the prosecution of the butcher, it was reported that the firm admitted that work surfaces, knives and other equipment at the shop were not kept clean and, as a result, food was not protected against contamination, rendering it unfit for human consumption and injurious to health (English, 1998). It was also reported that a boiler which was used to cook joints of meat and a vacuum packing machine used for packing both raw and cooked products were contaminated with *E. coli* O157.

The controls required to prevent contamination have already been described. However, for retail products sold from the butchery counter or from shops a further potential route of contamination arises during retail display and handling of the product from the display counter. Any ready-to-eat product is susceptible to contamination when stored in an open condition in a display cabinet prior to being handled for slicing or weighing and packing. In an environment where raw meat is also present, as in butcher shops, the possibilities for contamination are much increased because of the likely presence of enteric pathogens on the raw meat. The principles of good personal hygiene, segregation of raw and cooked meats and effective cleaning procedures are essential in preventing outbreaks.

Training of staff in all aspects of food hygiene is essential. Staff who have a good understanding of basic food safety and how their actions can compromise safety are far better equipped to manage the hazards around them, whether in manufacturing, retailing or even catering environments, than those who are merely following prescriptive work instructions.

METTWURST: AUSTRALIA

South Australia experienced a large outbreak of *E. coli* food poisoning between December 1994 and February 1995 (Cameron *et al.*, 1995b). A total of 23 cases of HUS were reported among children under 16 years of age, primarily in the Adelaide region of South Australia (Cameron *et al.*, 1995a). In the same region, doctors reported 33 further individuals suffering bloody diarrhoea or TTP where no other bacterial pathogen could be isolated from the stool specimens, together with 105 other individuals with gastrointestinal illness other than bloody diarrhoea, 30% of whom had a history of consuming the implicated product. One four-year-old girl

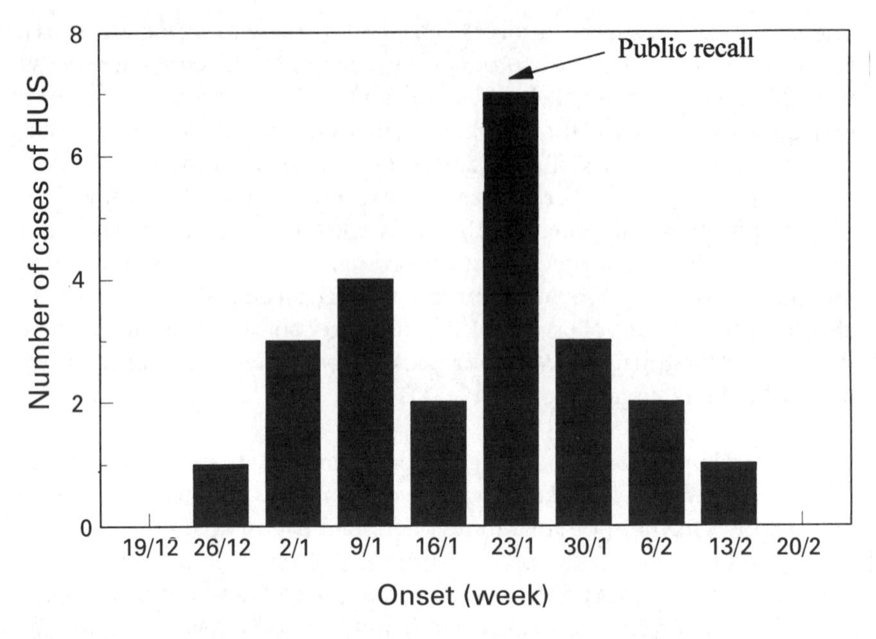

Figure 2.1 *E. coli* O111 outbreak associated with Mettwurst.

died in the outbreak. *E. coli* O111:NM (non-motile) was isolated from stool specimens of 16 persons with HUS, 1 person with bloody diarrhoea and 2 persons with gastrointestinal illness, although many more of the stool specimens from those affected were reported to be positive for Shigella-like toxin genes 1 and 2 (SLT 1 and SLT 2) using the more sensitive polymerase chain reaction (PCR) test (Cameron *et al.*, 1995a). A high proportion of those suffering HUS (16/23) reported consuming Mettwurst from a single manufacturer (Cameron *et al.*, 1995a). *E. coli* O111:NM was isolated from 4/10 samples of Mettwurst retrieved from HUS patients' homes. In addition, the strain was detected in 3/47 samples from retail stores and from the homes of those patients who had not suffered HUS but experienced diarrhoeal illness. Samples of sausage from other manufacturers taken from the homes of those suffering just diarrhoeal illness, together with samples from retail stores, did not reveal any *E. coli* O111:NM (Cameron *et al.*, 1995a). The outbreak received significant media attention (Sumner, 1995) and a public recall of the Mettwurst resulted in the rapid reduction of further infections (Figure 2.1).

The implicated product was Mettwurst (Table 2.2) produced by a small goods manufacturer. Mettwurst is an example of a raw fermented meat product similar to a salami. The exact nature of the raw meat mix used in the manufacture of the implicated Mettwurst is not reported but

Table 2.2 Outbreak overview: Mettwurst

Product type: Raw fermented meat product
Year: 1994–95
Country: Australia
Levels: Present in finished products but levels not reported

Possible reasons
(i) Contamination of raw meat mix
(ii) Inadequate process control
(iii) Cross-contamination to finished product

Control options*
(i) Slaughterhouse standards monitored as part of supplier quality assurance programme
(ii) Regular monitoring of raw meat intake for indicators of hygienic processing
(iii) Monitoring of process changes, e.g. pH, moisture content, etc.
(iv) Training of staff in safe food hygiene practices
(v) Effective segregation of raw and ready-to-eat meat products during manufacture and retail display
(vi) Operation of effective cleaning procedures

* Suggested controls are for guidance only and may not be appropriate for individual circumstances. It is recommended that proper hazard analysis is carried out for every process and product to identify where controls must be implemented to minimize the hazard from *E. coli*.

fermented meats can be manufactured from raw pork, raw beef or a mixture of the two. The implicated manufacturer is reported to have held raw beef, pork and lamb as raw materials for a variety of further processed products (Anon., 1995c). A typical fermented meat production process would begin with the raw meat being chopped into a fine paste and mixed with bacterial starter cultures and other ingredients such as salt, herbs and spices. The implicated Mettwurst included garlic in the raw ingredients but the process did not involve the use of starter bacteria. Like salami, the product is fermented at ambient temperatures to reduce the pH and then dried to reduce the moisture and water activity. The product may be sold as a whole sausage product to consumers or it may be sliced for sale on delicatessen counters.

The exact reasons for this outbreak are not clear but with a process involving the manufacture of a raw fermented meat two likely scenarios are possible that could allow contaminating enteric pathogens to be present in the finished product: either they were present in the raw materials and survived the process or they were introduced as post-process contaminants during manufacture or retailing. Cross-contamination during retailing is an unlikely possibility as under such circumstances the outbreak would be expected to arise from contamination at a single retail outlet and published reports give no indication that all the affected individuals purchased the implicated product from the same retail store. The most probable route of contamination would have been through contamination of the raw material meats and organism survival of the process or cross-contamination during the manufacturing process. Raw meat will inevitably introduce enteric pathogens into the raw Mettwurst mix and although the levels of contamination are likely to have been low, the efficacy of the subsequent fermentation and drying process is critical to ensuring destruction of the organism. However, as the ability of the subsequent process to reduce levels of contamination is dependent on initial contamination levels, it is critical to ensure suppliers of raw meat employ methods to limit contamination of incoming raw meat with enteric pathogens. Unfortunately, small manufacturers are unlikely to have the technical resources to enforce the requisite standards in slaughterhouse hygiene at their suppliers and are often dependent on the abattoirs themselves and any local government inspection systems to ensure high standards of raw meat production. The standards of the abattoir supplying the meat to the implicated supplier are not reported but contamination must be expected with any raw incoming meat. One approach open to the manufacturer is to assess the hygienic status of raw meat by introducing regular monitoring for indicators of faecal contamination so that trends can be identified and adverse results raised with raw meat suppliers. Assuming that contamination was present in the raw meat, then a salami-type process will distribute contaminants throughout the batch during bowl chopping of the raw meats with other ingredients. It is obviously not possible to prevent this occurring but, to reduce the potential for subsequent batches to become contaminated, it is essential that high standards of cleaning are employed following each production batch. When trying to achieve a reduction in the number of contaminating enteric pathogens, the most critical stages of the process are fermentation and drying. Following a similar outbreak in the USA caused by *E. coli* O157:H7 contamination of a fermented raw meat product (Alexander *et al.*, 1995), which resulted in over 20 illnesses, a large amount of work was conducted to determine the factors contributing to the safe manufacture of these products. Among the most important were starter culture activity to ensure rapid growth with concomitant production of organic acids during the early stages of fermentation and a low water activity brought about by extended drying times. To facilitate rapid starter culture growth, warmer fermentation conditions were believed to be advantageous. The outbreak strain in the USA was reported to be acid tolerant, but the recommendations concerning active fermentation and drying for the reduction of *E. coli* O157:H7 would be anticipated to achieve similar reductions in contaminating enteric pathogens such as the *E. coli* O111:NM implicated in the Mettwurst outbreak. The company manufacturing the implicated Mettwurst did not use starter cultures, a practice that had been discontinued as it was believed that little difference existed between products manufactured with or without starter bacteria. However, this contravened local legislation as a change in the Food Standards Code in 1985 had made it compulsory to include the use of starter bacteria. It is also clear from the coroner's report (Anon., 1995c) that process controls were not in place or operated effectively and critical control points were reportedly not monitored. Indeed, process control systems such as monitoring of pH decrease during fermentation were not carried out during the period of manufacture of the implicated Mettwurst. It is concerning to note that such process controls had been highlighted as being absolutely essential to the implicated manufacturer following a previous salmonellosis outbreak that implicated another fermented meat product made by the same manufacturer several years earlier. Inadequate understanding of the hazards inherent in this product was further demonstrated in the coroner's report, which highlighted that the person most intimately connected with the process 'did not understand the significance of water activity or content as a safety issue, including its importance in relation to the growth of micro-organisms'. Product was reported to have been regularly released before it was fully matured and was sometimes sent out only two or three days after it had been made, which again reinforces a lack of understanding about the importance of drying during maturation in the control of enteric pathogens. Clearly, any manufacturer of raw fermented meat products should recognize that fermentation and drying times and temperature are critical to the safety of the product, and consequently they have a responsibility to monitor these changes and ensure they occur correctly and consistently from batch to batch. In addition, to assess the efficacy of their specific process it would be prudent to determine the effect of the process on the reduction of enteric pathogens such as Vero cytotoxin-producing *E. coli* (under carefully controlled experimental conditions at suitable research facilities). If the process is found to result in insufficient reduction, then a process modification may need to be introduced, e.g. elevation of fermentation temperatures. Whilst it is most likely that contamination by the organism occurred in the production of the raw material and the organism then survived the process, it is also possible that contamination could occur following the fermentation and drying stages. Vero cytotoxin-producing *E. coli* could be introduced due to cross-contamination of finished product with raw meat or as a result of handling by personnel during packing. Indeed, in some outbreaks it has been postulated that cases may have occurred because of cross-contamination of products in retail display cabinets or during subsequent slicing. Prevention of cross-contamination from raw meats to processed ready-to-eat foods can only be achieved by effective segregation procedures during the manufacture and retailing operations. In addition, contamination from personnel will be controlled by adherence to good basic standards of food hygiene, which comes from effective training of staff.

Following the widespread public concern in Australia after the Mettwurst outbreak, the Australia New Zealand Food Authority (ANZFA) introduced specific requirements for manufacturers of fermented comminuted meat products (Australia New Zealand Food Authority, 1996a) and a new national food hygiene system (Australia New Zealand Food Authority, 1996b). The code required salami and Mettwurst manufacturers to comply with the following criteria:

- storage temperature for raw meat must be 5°C or less
- a starter culture must be used for fermented meat manufacture
- use of batches of out-of-code product or other waste material is prohibited
- monitoring of the raw meat and the finished product must be conducted for *E. coli* and records retained
- the pH of the product and the temperature in the fermentation room must be monitored
- the manufacturing process must be capable of delivering a 3 log reduction in *E. coli*
- the final product must have no *E. coli* in at least four out of five 10 g sample units
- all records must be kept for two years.

From an examination of outbreaks of VTEC in traditional products like fermented meats it is clear that the organisms appear to be selectively favoured because of their ability to survive at low pH, a preservation factor essential to the safety of such products. Because the infective dose of some serotypes is low, the survival of even a few cells may be enough to cause illness. In addition, the increasing occurrence of the hazard in raw meat will inevitably lead to more outbreaks in the future as a greater incidence in raw materials will further expose the frailty of some processes that are not capable of achieving sufficient reduction in the initial contaminating levels. Understanding the factors contributing to the safety of a process should be paramount for all manufacturers of raw fermented meat products as the long-term future of these traditional products relies on the avoidance of future outbreaks.

UNPASTEURIZED APPLE JUICE: USA AND CANADA

A number of outbreaks of *E. coli* O157 infection have occurred in recent years in which the consumption of apple juice has been implicated (Table 2.3). In 1991 an outbreak of haemorrhagic colitis and HUS was reported in south-eastern Massachusetts, USA due to consumption of fresh apple 'cider'. 23 people suffered illness and the causative organism was *E. coli* O157:H7 (Besser *et al.*, 1993). In Canada, an outbreak of haemolytic uraemic syndrome was attributed to the consumption of fresh apple juice (Steele *et al.*, 1982). *E. coli* O157 was not isolated from the patients or the product but was suspected to have been the cause of the outbreak because of the symptoms of patients, which included bloody diarrhoea, abdominal cramps and haemolytic uraemic syndrome. A total of 14 cases was reported in this outbreak and, although the organism was not isolated from any product, *E. coli* (not O157) was isolated from some samples. Nevertheless, apple juice was implicated as it was a common foodstuff consumed by patients suffering illness. Two recent outbreaks of *E. coli*

Table 2.3 Outbreak overview: apple juice

Product type: Unpasteurized, unfermented, fresh-pressed apple juice
Year: 1980, 1991 and 1996
Country: USA and Canada
Levels: Levels not reported

Possible reasons
(i) Use of fallen apples
(ii) Contamination of apples with cattle manure, resulting from its use as a fertilizer, or access to orchards of grazing cattle
(iii) Inadequate washing of apples
(iv) Survival of *E. coli* O157 in apple juice at low pH and chilled temperature

Control options*
(i) Fallen apples should not be used for unpasteurized juices
(ii) Prohibit use of untreated animal manure as fertilizer
(iii) Prevent access to orchards by grazing cattle
(iv) Wash apples in chlorinated water with free chlorine levels maintained
(v) Addition of preservative factors such as sodium benzoate (0.1%) may achieve faster reduction in contaminants

* Suggested controls are for guidance only and may not be appropriate for individual circumstances. It is recommended that proper hazard analysis is carried out for every process and product to identify where controls must be implemented to minimize the hazard from *E. coli*.

O157 infection associated with the consumption of fresh-pressed apple juice occurred in the USA in 1996. In the western USA commercial apple juice resulted in 66 cases and one death while in Connecticut, USA 14 people with confirmed illness caused by *E. coli* O157 had all consumed the same brand of apple 'cider' (Mshar *et al.*, 1997). In the latter outbreak, seven patients were hospitalized although no deaths have been reported. It appears from most of the published reports that the organism was not isolated from product, although products from the implicated batches were not always available for analysis.

Although the apple products implicated in these outbreaks have been called apple juice or apple 'cider', the terms are used to describe unpasteurized, unfermented, fresh-pressed apple juice. In the UK it is common to differentiate fermented and unfermented apple juice as cider and juice, respectively, whereas in the USA it is common to differentiate pasteurized and unpasteurized apple juice as juice and cider, respectively. Fresh-pressed apple juice is manufactured by collecting apples from the orchard, washing to remove debris, pressing to extract juice, adding preservative, where appropriate, e.g. potassium sorbate, and then filling into containers. Juices may be filtered to remove apple pulp but many fresh apple juices have a significant amount of suspended solids remaining in the product. Although all the details relating to the manufacturing process of the products in these outbreaks are not reported, it is clear that a common suspected origin of the *E. coli* O157 contaminant was the apples collected from the orchard. In both the Massachusetts and Connecticut outbreaks it is reported that fallen 'drop' apples were suspected to have been the source of the contamination. Contamination is believed to have arisen from the apples falling onto ground already contaminated with cattle faeces or which subsequently became contaminated with *E. coli* from the faeces of nearby cattle. Once contaminated with *E. coli* O157 the process of washing the apples, if employed, may help to remove some contaminants. This will be dependent on factors such as whether chlorinated water or just potable water was used, the time and temperature of contact with chlorinated water, and the ability of the washing process to remove adhering contamination on the apple, i.e. dry faecal matter. In the Massachusetts outbreak the apples were believed to include fallen fruit that were neither washed nor brushed prior to pressing (Besser *et al.*, 1993). After pressing, the products are usually chilled and retailed with a relatively short shelf life of 6 to 10 days. The shelf life of these products is restricted by the inevitable spoilage caused by yeasts, which are introduced on the fruit or during processing and selectively favoured at product pH of 3.5–4.0. Some fresh-pressed apple juice has preservative, such as potassium sorbate, added to inhibit yeast growth and extend shelf life.

Preventing outbreaks associated with products such as unpasteurized apple juice should be quite straightforward. Contaminated apples must be prevented from entering the process in the first place. The use of fallen apples will inevitably lead to contaminants from the soil being transferred to apples. If the soil itself is contaminated with faecal pathogens from cattle manure spread for fertilization of the soil or from uncontrolled entry of grazing cattle, as evidenced by the recorded outbreaks, then the consequences for consumers from the subsequent product can be serious. Clearly the best option is to prohibit the use of fallen apples for the production of unpasteurized fruit juices. In fact, fallen apples are more likely to be of poorer quality due to damage, bruising or insect attack and are therefore likely to introduce spoilage microorganisms such as yeasts as well as pathogens. Use of such apples is a self-defeating practice as the greater the proportion of fruit contaminated with yeasts, the greater and faster the spoilage of the product. However, even if non-fallen fruit is used it is possible, although less likely, for contaminants to enter the process as a result of soil, dust and airborne contamination reaching the apple on the tree. Containers used in harvesting may also add contaminants. Reducing contamination sources in the orchard itself can be achieved by preventing access to cattle and other animals and prohibiting the use of untreated animal manure as a source of fertilizer. Composted manure (which if composted for several months should reach high enough temperatures to achieve some reduction in levels of vegetative pathogens) or artificial fertilizers can be used instead. Contamination of picked fruit can be reduced by effective washing processes. Ineffective washing, however, can actually serve to distribute local pockets of contamination from a few fruit to an entire batch via the wash water. It is therefore essential that any wash water used for apples is chlorinated, with sufficient free chlorine available to inactivate pathogens washed off the apple surface the water. In the Connecticut outbreak, the apples received from six orchards were reported to be washed and brushed prior to being pressed, although use of chlorinated water is not noted. Once introduced into the pressed fruit it has been historically considered that enteric pathogens would die out due to the extremely low pH (<pH 4) and the high acidity level. However, from the evidence of the apple juice outbreaks it is clear that *E. coli* O157 can, and does, survive for extended periods in such inhospitable environments. When inoculated into apple juice, levels of an *E. coli* O157:H7 strain, isolated from a patient suffering illness, remained constant at 8°C for up to 12 days and then gradually declined by 2 to 5 log cfu/ml within 20 days (Zhao *et al.*, 1993). At 25°C, however, the organism survived for 2 to 3 days and then decreased by >5 log cfu/ml within 3 to 6 days. The ability of *E. coli* O157 to survive in apple juice at 20°C was reported by Semanchek and Golden

(1996). Clearly, with apple juice it is not possible to rely on acidity and pH alone to ensure safety, and raw material control must be considered a critical control point. Some processors have resorted to the use of added preservatives to enhance the safety of apple juice. Zhao *et al.* (1993) reported a minimal effect of potassium sorbate on *E. coli* O157 survival in apple juice, survivors being detected up to 20 days. In contrast, potassium sorbate (0.1%) combined with potassium benzoate (0.1%) was shown to reduce the survivors in apple juice to a less than detectable level within 7 days. Products such as freshly squeezed juices have long had excellent customer appeal and an associated perceived health benefit but the ability to ensure the safety of these products is critical to maintaining consumer confidence in a product group that has now repeatedly been implicated in outbreaks of *E. coli* O157 infection. As with many products, it is the application of basic principles of hazard analysis and hygiene that can maintain the safety of these products.

PASTEURIZED MILK: SCOTLAND

In May 1994, Scotland experienced a large outbreak of *E. coli* O157 infections with over 100 people reported to be affected (Upton and Coia, 1994). The outbreak occurred in West Lothian and the majority of those affected were young children (Sharp *et al.*, 1995). The organism responsible was *E. coli* O157:H7 phage type 2, VT2, which was isolated from 69 of the patients. Nearly one-third of those affected required hospitalization. Nine children aged between 11 months and 11 years developed HUS and one elderly women developed TTP. The outbreak was attributed to pasteurized milk from a single dairy (Table 2.4). It is important to note that direct culture methods failed to detect the *E. coli* isolate from several clinical and dairy environmental samples but, when methods were supplemented with the immunomagnetic separation technique (Wright *et al.*, 1994), a number of isolates were recovered (Upton and Coia, 1994). *E. coli* O157 was isolated from dairy samples, including a pipe transferring milk from the pasteurizer to the bottling machine, a discarded bottling machine rubber seal and a bulk milk tanker that contained raw milk from one of the supplier dairy farms. The organism was also subsequently isolated from bovine faecal isolates taken from one of the supplier farms (Coia and Hanson, 1997). All isolates were considered to be indistinguishable from those isolated from the clinical cases based on phage typing and pulsed-field gel electrophoresis (PFGE) (Coia and Hanson, 1997).

It is not clear from the reports whether this outbreak occurred due to improper pasteurization of the raw milk or due to post-process

Table 2.4 Outbreak overview: pasteurized milk

Product type: Fresh pasteurized liquid milk
Year: 1994
Country: Scotland
Levels: Present in environmental samples from plant

Possible reasons
(i) Use of contaminated raw milk
(ii) Inadequate pasteurization of milk
(iii) Cross-contamination of raw to pasteurized milk via product, plant or personnel

Control options*
(i) Hygiene in milking parlour (udder, milking equipment, pipework and storage tanks) and during distribution (milk tankers)
(ii) Process controls/monitors on pasteurizer, e.g. divert valve, thermograph
(iii) Segregation of raw and pasteurized product
(iv) Implementation and maintenance of procedures to avoid post-pasteurization contamination, i.e. personnel practices
(v) Operation of appropriate cleaning schedules and associated monitors of cleaning efficacy, including ATP bioluminescence hygiene monitoring

* Suggested controls are for guidance only and may not be appropriate for individual circumstances. It is recommended that proper hazard analysis is carried out for every process and product to identify where controls must be implemented to minimize the hazard from *E. coli*.

contamination. However, as the *E. coli* O157 isolate was found in the faeces of a bovine herd supplying milk, in a bulk milk tanker and in samples taken post pasteurization, it is clear that raw or improperly pasteurized milk must have gained access to the plant either by inadequate heat treatment or by ineffective segregation of raw and pasteurized milk in the transfer and handling systems. The production of pasteurized milk is a fairly simple process starting with the receipt of milk, usually in bulk milk tankers, from a number of dairy farms. The milk is stored in silos and, where necessary, it is separated into skimmed milk and cream. Milk is usually standardized to achieve the required fat content prior to pasteurization. It then usually passes to a buffer tank prior to filling into cartons or bottles. The milk is chilled to <5°C and is usually allocated a 6 to 10 day shelf life.

The exact process involved in the manufacture of the implicated milk is not reported but clearly the most critical stage in the pasteurization of milk is the heat process itself. Milk is known to contain, from time to time, a range of different microbial pathogens and probably the most frequently occurring are the enteric bacterial pathogens. The primary route of microbial contaminants into raw milk is likely to be from the udder and milking equipment, which may be contaminated with faecal material that then passes into the milk. As pathogens including *E. coli* O157 have been isolated from the faeces of dairy cattle it is clearly possible for them to pass into the milk supply. Mechie *et al.* (1997) investigated the incidence of *E. coli* O157:H7 in dairy cattle associated with an outbreak of human infection due to consumption of unpasteurized milk from a herd in the Sheffield area (UK) during 1993/4. *E. coli* O157:H7 was isolated from 153 out of 3593 rectal swabs taken from cows, heifers and calves in a 15-month period. In the same study *E. coli* O157:H7 was isolated from milk samples taken from two animals (one from a milk jar and one from fore-milk). In the UK, milk is pasteurized at a minimum of 71.7°C for 15 s and, providing this is carried out effectively, vegetative pathogens including *E. coli* O157 will be destroyed. To avoid underpasteurization of milk it is essential to have systems in operation that monitor the heat process given to the raw milk. In most commercial milk pasteurization plants it is usual to have in-line sensors that continuously monitor the temperature of pasteurization. These are connected to automatic valves which, when the temperature drops below a pre-set value, re-route the unpasteurized milk into a divert tank so as to prevent unpasteurized milk passing into the finished product buffer tank and on to the filling operation. Divert valves must be checked to ensure they operate correctly and temperature monitoring equipment must be calibrated regularly. Such automated systems should be supplemented with continuous temperature chart recorders, which can be inspected after each operation. The efficacy of the milk pasteurization process is often checked with a rapid test on the pasteurized bulk milk using a phosphatase test. Pasteurization inactivates the phosphatase enzyme therefore high levels of this enzyme detected post pasteurization provide an indication that some unpasteurized milk may be present. In smaller milk-pasteurization plants sophisticated equipment may be lacking; pasteurization may involve heating bulk milk in a single jacketed vessel to achieve temperatures in excess of 63°C for 30 min. Process control may also be minimal with checks of temperature relying on visual assessment rather than automated equipment with chart recorders and divert valves. In such circumstances, it is possible to envisage circumstances where raw milk may not be processed adequately due to insufficient attention being paid to monitoring the time and temperature of the process. However, even with sophisticated machinery it is equally possible for failures to occur as a result of divert valves being manually overridden by operators who are insufficiently trained and are unaware of the hazards presented by such acts. The most usual cause of pathogens being present in pasteurized milk is therefore inadequate pasteurization, and considerable attention must be given to the systems and process controls in place at any milk-pasteurization facility to prevent this most basic of failures from occurring. The second most likely reason for contamination of pasteurized milk is post heat-process contamination with pathogens, caused by improper valve operation after pasteurization, resulting in raw milk mixing with pasteurized milk. Such occurrences arise either because of poor design of the process equipment or because of human error in routing products incorrectly. The design of equipment and operating practices based on a hazard analysis approach can assist in preventing such occurrences. Coupled with operating systems that, through the training of staff, help operatives to conduct their jobs effectively, serious food poisoning incidents involving pasteurized milk can be eliminated. However, the potential for more simple things to contribute to post-process contamination must not be overlooked. Dairies rarely have complete segregation of raw-milk areas from pasteurized-milk areas. Many operators are able to pass into the post-pasteurization plant (usually filling rooms) without the normal high/low risk separation procedures employed in factories manufacturing heat-processed ready-to-eat products from raw materials potentially contaminated with enteric pathogens. The potential for contamination of milk during filling is naturally limited, but personnel entering such areas from the tanker reception area or raw-milk handling areas are likely to bring with them contaminated footwear or other clothing. Such practices are more common with small dairy plants where the number of staff is limited. Attention to the need for clothing changes and disinfection of footwear must be given high priority to help prevent equipment in the post-pasteurization area becoming contaminated from factory personnel.

The pasteurization of raw milk to produce pasteurized milk should be one of the most simple of processes to control; raw milk hygiene, effective heat processing and avoidance of post-process contamination are the keys to success. However, without sufficient attention to the application of basic hygienic design and operation in raw milk collection, equipment function and processing, incidents of food poisoning from even this most basic of foods will continue to burden society.

SOFT RIPENED FRENCH CHEESE: USA

The first fully documented foodborne outbreak due to enteropathogenic *E. coli* occurred in the USA in 1971 (Marier *et al.*, 1973). In this outbreak over 387 people were reported to have suffered gastrointestinal illness between October and December 1971. The most commonly reported symptoms included diarrhoea, fever and nausea, although some persons also suffered bloody diarrhoea. On average, the duration before onset of symptoms was 18 h although this varied from 2 to 48 h. The average age of those affected was 34. 12 patients required hospitalization although no deaths were reported. Outbreaks of infection occurred in a total of 14 different states across the country, from Connecticut to California. The organism responsible for the outbreak was enteropathogenic *E. coli* O124:B17. This isolate was reported to be invasive but not enterotoxigenic according to the tests available in 1971 (Marier *et al.*, 1973). The implicated food was mould-ripened soft cheese imported from a single manufacturing plant in France (Table 2.5). The cheese types implicated included Brie, Camembert and Coulommiers, reported to be from identical manufacturing processes with differences only in product size and labelling. One dealer, keen to prove the safety of the cheese, was reported to have consumed some implicated cheese and was subsequently added to the list of those affected in the outbreak. A total of 71 cheese samples were taken for analysis, representing 13 production days

Table 2.5 Outbreak overview: ripened soft cheese

Product type:	Mould-ripened soft French cheese (Brie, Camembert)
Year:	1971
Country:	USA
Levels:	10^5–10^7 per gram *E. coli* O124:B17

Possible reasons

(i) Use of unpasteurized milk
(ii) Contamination of production equipment with contaminated river water
(iii) Survival and growth of *E. coli* O124 in mould-ripened product

Control options*

(i) Use of pasteurized milk
(ii) Use of treated mains water for cleaning purposes
(iii) Treatment of non-mains water with chlorine to destroy enteric pathogens
(iv) Microbiological monitoring of water supplies for cleaning purposes
(v) Operation of apprioriate cleaning schedules and associated monitors of cleaning efficacy, including ATP bioluminescence hygiene monitoring

* Suggested controls are for guidance only and may not be appropriate for individual circumstances. It is recommended that proper hazard analysis is carried out for every process and product to identify where controls must be implemented to minimize the hazard from *E. coli*.

from the implicated manufacturing plant. *E. coli* O124 was isolated from 25 of the samples. All cases of illness, however, related to cheese produced in a 2-day period; 45 of the 71 samples taken were from these two production days, of these, 23 yielded *E. coli* O124. It is interesting to note that *E. coli* O125 was also isolated from 2/71 samples of product, although this strain was not isolated from any of the stool specimens. In the cheeses found to contain the outbreak strain the levels of *E. coli* O124 were reported to range from 10^5 to 10^7 per gram.

The published reports do not clarify whether the cheeses implicated in this outbreak were manufactured from pasteurized or raw milk. The three types of cheese, Brie, Camembert and Coulommiers, are all essentially identical in manufacture and differ primarily in the size of the cheese round. The cheeses are manufactured by adding lactic acid starter culture bacteria to warmed pasteurized or unpasteurized milk. The milk is fermented for a short period and then rennet is added to coagulate the proteins and time is allowed for separation of the whey from the curd. This is assisted by cutting the curd into cubes prior to drainage. The curd is then scooped into the cheese mould and placed on racks at ambient temperature for 1 to 2 days to allow further drainage of the curd. After this, the cheese is usually removed from the mould and dry salted by rubbing salt over the surface to give a fine covering. The cheese is then inoculated with a spray of mould spores, which develop during ripening to produce a mould coat on the surface. The product is usually ripened for several weeks at temperatures between 8 and 12°C prior to packing, chilling and distribution. It may be sold as whole cheese or cut into wedges and wrapped at the factory or in the retail store. The cheeses have a shelf life of several weeks at chill temperatures. The finished product has a characteristic white mould covering and a soft texture, which becomes 'runny' towards the end of life due to the action of microbial enzymes during the shelf life. The investigation conducted by the French authorities following this outbreak found *E. coli* O124 in a curdling tank (fermentation tank) and also in finished products. No major process deficiencies were revealed although a failure in the equipment used to filter river water was noted during periods that coincided with production of cheese implicated in the outbreak. The river water was used for cleaning purposes in the factory.

In assessing the lessons that could be learnt from an outbreak such as this it is clear that much rests on whether or not the raw milk was subjected to pasteurization. Raw milk is known to be a source of a variety of pathogens, including VTEC, and without a pasteurization stage some pathogens will inevitably enter the process. With high standards of animal husbandry and hygiene during milking and storage of the milk at the farm and during transport, the incidence and levels of contamination in raw milk can be minimized but contamination will still arise occasionally. Pasteurization has been recognized for many years as a successful way of ensuring the safety of milk because it achieves a high level of destruction of a variety of vegetative bacterial pathogens, including enteric pathogens. The process involved in the manufacture of a mould-ripened soft cheese made from raw milk is highly unlikely to result in the destruction of enteric pathogens. Although the initial stages of bacterial starter culture fermentation do achieve a reduction in pH from near neutral to pH <5.0, together with the concomitant production of organic acids, primarily lactic acid, such changes are more likely to merely restrict the growth of pathogens like VTEC rather than destroy them. Studies of *E. coli* O157 survival during the manufacture of cottage and cheddar cheese actually demonstrated a 1–2 log increase during cheese fermentation and prior to ripening (Arocha *et al.*, 1992; Reitsma and Henning, 1996). In addition, during the ripening stages (8–12°C) the mould growth on the surface markedly changes the pH at the surface from pH <5.0 to near neutral, which again favours the survival and even growth of some enteric pathogens. Therefore, if the cheeses were made from raw milk it is clear that the potential for outbreaks like this will remain. However, from the information given by Marier *et al.* (1973) it appears probable that this outbreak is likely to have been caused by post-process contamination. Mould-ripened cheeses are equally likely to cause outbreaks of foodborne illness whether they are made from raw or pasteurized milk if inadequate standards are in operation to control post-process contamination. The use of river water in the implicated plant to clean equipment is not an acceptable practice in the food manufacturing industry. Although it is reported that the river water was subject to filtration and the filtration apparatus was functioning correctly at the time that the implicated cheeses were produced, it is usual to clean equipment with treated mains water. Surface water, such as that from rivers, is subject to contamination from agricultural run-off or even sewage discharge, and its use in a food factory brings with it a range of microbial hazards. As water is used to both clean and rinse surfaces and comes into contact with most product contact surfaces, the potential for contamination must be recognized. The quality of the water supply is critical to the safety of most food products that are destined to be consumed without further processing. In the UK the quality of the water supply is the responsibility of the local water company up to the point it is received onto the manufacturer's premises. After this point, it becomes the responsibility of the manufacturer to maintain the integrity and safety of storage tanks and pipework used to hold and carry the water. It is usual to check water quality at regular intervals by monitoring the level of microbial contaminants, including both total bacterial counts and indicators of faecal contamination such as coliforms and/or *E. coli*. Such monitoring also takes place for private deep borehole supplies and it is usual to treat the borehole water on site by chlorination prior to use in factories. In addition to water, the possibility of contamination of the product by other forms of post-processing contamination should not be overlooked as contamination with certain other bacterial pathogens during mould ripening is unlikely to diminish because the conditions during ripening are suitable for microbial survival and even growth. High standards of equipment and environmental hygiene resulting from the operation of effective cleaning and disinfection programmes in all manufacturing areas will help to achieve control of such contaminants. In addition, adequate personal hygiene practices are essential for operatives handling products that are destined to be consumed without further bactericidal processing, although failure in this area was not reported as being contributory to this particular outbreak.

BEEFBURGERS: USA

One of the largest outbreaks of *E. coli* infection to have occurred was recorded in the USA between November 1992 and February 1993. This multi-state outbreak in California, Idaho, Nevada and Washington resulted in over 700 reported cases with over 500 laboratory-confirmed infections and four deaths (Davis *et al.*, 1993). In total, 732 people were affected, resulting in the hospitalization of 195 of them. 55 patients subsequently developed HUS or TTP and of the 614 cases in Washington state over 10% were considered to represent secondary cases, becoming infected from cases who had previously contracted illness from the original food source. The age of patients ranged from 0 to 74 years, although the median age range was 7.5 years. The strain isolated in this outbreak was *E. coli* O157:H7 (VT1+, VT2+). The product implicated as the source was identified as beefburger patties sold by a single restaurant chain (Table 2.6). Minced beef implicated in the outbreak was tested microbiologically by a number of laboratories using several different methods (Advisory Committee on the Microbiological Safety of Food, 1995) and *E. coli* O157:H7 was isolated from the samples with the same lambda phage restriction fragment length polymorphism (RFLP) profile found in 61 out of 63 isolates recovered from infected individuals. From the experimental work conducted during this investigation it was possible to estimate that between 1 in 300 and 1 in 400 patties from the implicated source which were served resulted in illness (Advisory Committee on the Microbiological Safety of Food, 1995) and that contamination levels ranged from 1 to

Table 2.6 Outbreak overview: beefburgers

Product type:	Cooked beefburger patties from a restaurant
Year:	1992–93
Country:	USA
Levels:	1–15 per gram

Possible reasons

(i) Use of contaminated beefburger patties
(ii) Inadequate cooking of burgers in the restaurant
(iii) Survival of *E. coli* O157 in the cooked ready-to-eat burger

Control options*

(i) Supplier assurance programme
(ii) Procedure in place to ensure destruction of *E. coli* O157 during cooking, including defined process times and temperatures
(iii) Process verified using worst-case conditions (thickest burger, coldest burger, coldest position on cooking equipment)
(iv) Regular monitoring of process efficacy with temperature probing of cooked product during the day
(v) Staff practices to avoid cooked product recontamination
(vi) Public education to eat only fully cooked burgers

* Suggested controls are for guidance only and may not be appropriate for individual circumstances. It is recommended that proper hazard analysis is carried out for every process and product to identify where controls must be implemented to minimize the hazard from *E. coli*.

15 organisms per gram. With calculated levels of 40 to 600 organisms per raw beefburger, the levels remaining after cooking must have been very low and therefore, the infectious dose in cooked beefburgers must have been low. The outbreak is believed to have been caused by inadequate cooking of the beefburger patties, which allowed survival of contaminating *E. coli* O157 in the original burger.

This outbreak, although very large, represents only one of a number of outbreaks of food poisoning attributed to *E. coli* O157 that have reported links to the consumption of beefburgers or ground beef. In the UK in 1993 beefburgers from a local butcher's shop were implicated in a small community outbreak. In this case, indistinguishable strains of *E. coli* O157:H7 were isolated from raw meat and seven of the eight patients (Willshaw *et al.*, 1994). Wells *et al.* (1983) reported an association between beefburger sandwiches purchased from outlets of the same fast food chain and at least 47 cases of haemorrhagic colitis in Oregon and Michigan, USA in 1982. Belongia *et al.* (1991) reported on an outbreak of *E. coli* O157:H7 colitis associated with the consumption of pre-cooked beefburger patties affecting 32 students at a Minnesota junior high school.

All the outbreaks noted above have one thing in common: effective cooking of the products would probably have prevented them. The manufacturing processes involved in the production of raw minced beef and beefburgers are clearly not aseptic processes. Contamination of exposed meat surfaces with faecal matter during slaughter, evisceration, removal of the hide and further processing needs to be minimized. Adherence to high standards of slaughterhouse hygiene will reduce the frequency and perhaps the level of contamination but until processes such as steam pasteurization or other carcass decontamination techniques become common practice, contamination of raw beef products by enteric organisms has to be anticipated. The reason why beefburgers have been associated with such a high number of outbreaks is explained by understanding the processing involved. Beefburgers are manufactured by comminuting raw meat and mixing it with herbs, spices, and sometimes vegetables and flavouring other ingredients. Contamination of the animal carcass or further processed meat primals is usually restricted to external surfaces and is the result of faecal contamination during slaughter, carcass dressing and butchery or cross-contamination during further handling and processing from common production surfaces, utensils and personnel. Any contamination on the external surface of a whole joint or piece of meat, e.g. steak, is susceptible to the highest heat during cooking but with beefburgers any external contaminants on the raw meat become distributed throughout the product during comminution. Contaminating VTEC may therefore be present both on the surface and in the centre of the product. In addition, beefburgers are manufactured in different sizes (diameter and depth) and are often distributed frozen. Under such circumstances it is easy to envisage why beefburgers may be responsible for outbreaks of illness due to contamination by organisms capable of causing severe infections from a very low infectious dose.

Thick, frozen beefburgers, cooked from frozen, are known to be far more difficult to cook throughout, without burning the outside, and therefore undercooking of such burgers is probably more common than fresh meats. However, there is no clear evidence that the burgers involved in the above outbreaks were actually thick, frozen burgers. In fact, the explanation for some of the outbreaks appears much simpler than this. In the outbreak in Washington it is reported that when beefburgers were cooked according to the policy in place at the restaurant outlets, 10 out of 16 regular beefburgers had internal temperatures below 60°C (Bell *et al.*, 1994); this is insufficiently high to kill *E. coli* O157. In addition, prior to (and even after) these outbreaks fast food outlets in the USA commonly asked customers whether they required beefburgers cooked well done, medium or rare. This is an extension of the practice throughout the world for cooking and presenting beef steaks but microbiologically the risk is significantly different. A rare steak will still normally achieve sufficiently high temperatures at the surface to kill any pathogens present; contaminants in the rare part of the steak are likely to be minimal, if present at all. A rare burger, on the other hand, will inevitably harbour foodborne pathogens in the centre, which will remain unaffected by the mild temperatures achieved. Although the cooking procedure in the restaurant chain involved was clearly not capable of achieving a consistent temperature of 60°C, it is important to note that the minimum temperature required for these products in US law at the time of the outbreak was actually 60°C (Anon., 1993a), although Washington State law required temperatures of 68.3°C. The US Food and Drug Administration (FDA) Model Food Code provisions were reported to be based on the assumption of fairly low levels of contamination and it is clear that such a low temperature (60°C) would need to have been maintained for over 8 min to ensure a 5 log reduction in contaminating *E. coli* O157, whereas it is common practice in the retail industry not to hold beefburgers at high temperatures for extended periods but to use a flash heat process (Anon., 1993a). As a consequence of the outbreak, the FDA Model Food Code provisions were increased to 68.3°C for ground beef products to ensure a sufficient destruction of *E. coli* O157.

In reviewing this outbreak it is straightforward to describe measures that need to be implemented to avoid such occurrences. While it is important to ensure that any bacterial contamination of raw beefburger or ground meat products is kept to a minimum by operating vendor assurance schemes, the systems in place within a food service establishment must be focused on achieving two clear goals: first, to cook the product to a sufficient temperature to eliminate the hazard and, second, to avoid recontamination of the cooked product from the raw product or from contaminated people or equipment. As most of the above outbreaks have actually involved cooking issues these will be the focus for this study of the outbreak. Cooking a raw beefburger product in a food service environment or even in a factory requires exposing the product to a certain process time and temperature. As the temperature of each individual product after processing cannot easily be checked, the factors for ensuring effective cooking include:

- initial product temperature
- product characteristics (thickness, etc.)

- process temperature
- process time.

Such factors should be accounted for in any process validation before any process is implemented. In a food service environment it is usually possible to specify a product of a standard thickness, and process validation should be conducted on products likely to represent a worst case situation, i.e. the thickest. Variation in temperature across the cooking equipment needs to be established, again to assess the worst case situation. Once this has been done, the process validation is conducted by assessing the times necessary to reach the required internal temperatures that will achieve a sufficient reduction in contaminating microorganisms. Instructions to food service operators need to be clear concerning the use of chilled or frozen burgers, preheating times for the cooking equipment to achieve cooking temperatures and minimum times for cooking. Staff operating in such environments should be trained in basic food hygiene practices so that they have an understanding of the potential hazards involved in cross-contamination and also in the hazards of undercooking of the product and the ways in which undercooking could occur. Temperature checking should also form a regular part of the operational procedures throughout the day in a food service environment. Critically, this should include a temperature check of products on start-up, but clearly the focus must be on adherence to operating times and temperatures and sole reliance must not be placed on the temperature check. Although ensuring beefburgers are cooked thoroughly by manufacturers and food service operators should avoid recurrences of such outbreaks of foodborne illnesses, an essential element in the control of further incidents is public education of the hazards involved. For this purpose, cooking instructions on retail packs of raw beefburgers need to take account of the same considerations of product temperature, thickness and variation in cooking methods that the product may be exposed to in the consumer's kitchen so that the public has the necessary information to deal with potential hazards at home. In addition, the potential dangers of consuming rare (undercooked) beefburgers must be highlighted to the public to allow them to understand the inherent hazards associated with such practices. Indeed, in the USA a large public awareness campaign has now been implemented relating to the cooking of beefburgers (Anon., 1996a).

SPROUTING VEGETABLES: USA AND JAPAN

Between May and November 1996 Japan experienced the largest recorded outbreak of *E. coli* O157 infection. Reports vary, but it is estimated that more than 9500 people were involved in the outbreak (Gutierrez, 1996), which resulted in 10 deaths (Gutierrez and Netley, 1996). The strain was identified as *E. coli* O157:H7 (VT1 + VT2+) and although the outbreak occurred in more than 24 prefectures across the country (Gutierrez and Netley, 1996), over 6000 cases occurred in Sakai City, Osaka prefecture (Watanabe *et al.*, 1996). Although the same *E. coli* O157:H7 type appeared to be responsible for the outbreaks, Watanabe *et al.* (1996) could distinguish two different genotypes with the Sakai City strain differing from the strains isolated from other prefectures. The outbreak, which principally affected children, was attributed to school lunches. Radish sprouts present in the lunches were believed to be the source of the organism but *E. coli* O157 could not be isolated from water, seeds or stool tests of the employees at the sole implicated radish sprout producer (Gutierrez and Netley, 1996). Two outbreaks of *E. coli* O157 infection were reported in the USA between June and July 1997 (Como-Sabetti *et al.*, 1997). In Michigan, of 60 persons reporting with *E. coli* O157:H7 infection between June and July 1997 over 40 cases were found to be infected with the same strain of *E. coli* O157:H7 according to PFGE typing. The average age of those affected was 35 years (the age range was 2 to 79 years) and 96% reportedly suffered bloody diarrhoea with 54% requiring hospitalization. Two people developed HUS and one had TTP. No fatalities were reported. Over the same period in Virginia, 24 isolates from persons infected with *E. coli* O157:H7 were found to be identical by PFGE and, in the subsequent investigation, these isolates were found to be identical to the isolates from those infected in the Michigan outbreak. The strain was identified as *E. coli* O157:H7 PT32. In case control studies conducted independently following each outbreak, alfalfa sprouts were identified as the most likely source with 60 and 68% of those questioned in Michigan and Virginia, respectively, having consumed alfalfa sprouts compared to only 5 and 13% of the controls. In the Michigan outbreak all implicated seeds came from a single producer and the investigation found no evidence of unhygienic practices at the sprouting facility, with none of the environmental samples yielding positive *E. coli* O157. In the Virginia outbreak all implicated products were supplied by a single sprouting facility, which was different to that in the Michigan outbreak. Again, no evidence of poor hygienic practice was found in the ensuing investigation, with no positive results for *E. coli* O157 from environmental samples. In the full investigation of the two outbreaks one common factor was revealed: the use of a common source of seeds for the production of the alfalfa sprouts. The seeds were therefore considered to represent the source of the outbreak (Table 2.7).

Sprouting vegetables can be prepared in a variety of ways, usually involving germination of the seed at elevated temperatures around 30°C in

Table 2.7 Outbreak overview: sprouted vegetables

Product type:	(a) School lunches containing radish sprouts (Japan) (b) Alfalfa sprouts (USA)
Year:	1996
Country:	Japan and USA
Levels:	Not reported

Possible reasons

(i) Contamination of the seed
(ii) Survival of seed washing
(iii) Growth during seed germination
(iv) Distribution to vulnerable sector of community (Japan outbreak)

Control options*

(i) Control of agricultural practices that may contribute to hazard
(ii) Supplier quality assurance programme and intake screening of seed
(iii) Chlorinated wash/soak of seeds for extended time
(iv) Chlorinated wash of salad vegetables after germination

* Suggested controls are for guidance only and may not be appropriate for individual circumstances. It is recommended that proper hazard analysis is carried out for every process and product to identify where controls must be implemented to minimize the hazard from *E. coli*.

rooms of controlled temperature and humidity. Depending on the product, the seed usually germinates within several days and, after sprouting and some early growth, may be harvested immediately. Alternatively, the crop may be transferred outside to enclosed greenhouses to develop into a more fully grown product. These plants are usually watered during further growth by overhead sprays and are then harvested for sale. After harvesting, products may be washed in chlorinated water prior to spinning/draining and packing. The exact nature of the process will vary considerably depending on the product.

Although the investigations of both the Japanese and American outbreaks have not identified the specific cause of the outbreaks, it is clear that the key area of contamination for sprouting vegetables is the seed itself. The conditions of warm temperatures and high humidity during the sprouting process, which are critical to the development of the crop, are equally ideal conditions for the growth of enteric pathogens. Recent work investigating the growth of *E. coli* O157 on radish sprouts clearly demonstrated the growth potential of the organism during the process (Hara-Kudo *et al*, 1997): levels increased by 10 000-fold during the process (Jaquette *et al.*, 1996). Seeds used for the production of sprouting vegetables are often purchased on the open market from poorly developed countries. Under such circumstances it is clearly difficult to ascertain whether the methods of seed production and harvest were conducted in accordance with good agricultural practice. The use of animal manure as fertilizer, together with poor hygienic precautions by harvesting operatives, combine to compromise the safety of the raw material seed. Given the potential for the amplification of minimal contamination levels during the seed germination process, it is essential that the levels of such contamination in the raw material seed are extremely low and ideally absent altogether. Raw material assurance programmes, including inspection of the seed supply sites, would go some way to ensuring contamination is minimized. This should be supplemented with intake raw material tests of the seed to assess the degree of contamination using indicator tests for *E. coli* generally and supplemented with tests for specific enteric pathogens including *E. coli* O157 and *Salmonella* species. Of course, once purchased, it is equally essential to ensure that handling practices within the sprouting facility do not allow contamination to occur during handling of the seed and subsequent germination and growing processes. Chlorinated soaking of the seed will reduce levels of contaminants on the external surface of the seed; non-germinated seeds can tolerate levels in excess of 100 ppm free chlorine. Hygiene control of the germination environment is equally important in the prevention of product contamination. Depending on the seed type, the product may be sprouted using water or it may be grown on soil or, more usually, peat. In the latter case, the growing medium 'raw material' needs to be checked to ensure it is not contributing to the contamination of the product. Another major source of potential contamination is the water applied to the crop during growth. Many plants can tolerate fairly low levels of chlorine (<5 ppm) in the water and it may be appropriate to maintain a low residual chlorine level in the water to prevent excessive growth of any contaminating pathogens.

Products are usually harvested manually and in some cases they may be sold in the pot in which they were grown, e.g. mustard and cress, salad cress, rape and cress. The opportunity also exists for contamination of the product during harvesting and packing but any contaminants introduced at this stage are less likely to proliferate, although this will depend on the conditions prevailing in distribution and retail of the final product. Many products receive a chlorinated wash after harvest and prior to packing and this can assist in the reduction and control of pathogenic microorganisms. Critical to achieving successful effects is the use of a controlled chlorine dosing system for maintaining required levels of residual chlorine, e.g. >50 ppm, to both assist the product decontamination process and to control the levels of microorganisms in the wash water.

These products are usually sold as prepacks, held under refrigerated conditions and given a short shelf life of approximately four days, although in many countries products are also sold at ambient temperatures, e.g. stored loose on market stalls where the factors governing shelf life usually relate to visual deterioration of the product. Consumers are usually advised by most major retailers to wash produce of this nature prior to consumption although when presented as prepackaged units they are often sold as ready-to-eat. However, the degree to which a gentle washing by the consumer will reduce enteric pathogens, if present, is of considerable debate. Therefore, the safety of these products must lie with the control of the raw material seed itself, supplemented with control of the process conditions to ensure that any contaminant entering the process is kept to a minimum and the risk to human health reduced. Unfortunately, the nature of the process for these products will inevitably expose consumers to a higher incidence of *E. coli* O157 if the raw material seed growing and harvesting sites increase the use of natural fertilizers on the crops.

In addition, it is important to note that in the Japanese outbreak it is possible that the implicated radish sprouts may have resulted in only a restricted and small outbreak if the conditions of further processing, handling and distribution to children in their lunchboxes had been controlled to prevent further growth. Catering facilities should be operated in a way that ensures consistent application of high standards of hygiene to prevent cross-contamination from raw to ready-to-eat products, and from infected individuals, to products.

3

FACTORS AFFECTING THE GROWTH AND SURVIVAL OF *E. COLI*

GENERAL

It is clear that non-pathogenic and pathogenic types of *E. coli* are to be expected as contaminants of raw foods, particularly raw meats but also vegetable crops, which are at risk from contamination by warm-blooded animal faecal matter.

As it is impossible for many raw foods to be produced free from *E. coli* at source, it is important to establish animal husbandry and crop agricultural regimes that can make a positive contribution to minimizing the frequency and level of contamination of these primary raw materials by *E. coli*. The low infective dose of VTEC O157:H7 makes it particularly important to operate the highest possible standards of agricultural practice targeted to minimize direct faecal contamination of raw foods.

The introduction and consistent operation of animal husbandry and crop agricultural practices must be aimed at minimizing the faecal contamination load on raw material foods reaching food processing plants. Thereafter, raw material handling and processing procedures need to be structured and operated to minimize any multiplication of *E. coli* and prevent cross-contamination of the organism from animal or crop raw material foods to processed foods or equipment. This demands of the food processor a well-designed and reliably operated hygienic food production process based on hazard analysis critical control point (HACCP) assessments of each food process.

In addition to attention to the detail of cleaning and hygiene procedures, the treatment and formulation of food products are important for controlling any residual *E. coli* organisms and preventing their potential to cause harm to consumers. Within food production processes a variety of physico-chemical factors, used either singly or in combination, can be effective in controlling the survival and growth of *E. coli* during processing and also in the finished food products.

TEMPERATURE

Much of the early work carried out to determine the effects of temperature on the survival and growth of *E. coli* used mixtures of different types of *E. coli* (EPEC, ETEC, EIEC) (International Commission on Microbiological Specifications for Foods, 1996a) and it is only more recently that clearly targeted studies have been done to assess temperature effects on specific types of *E. coli*, such as VTEC O157:H7.

Table 3.1 shows the optimum and limits of temperature for the growth of *E. coli*, demonstrating the slightly narrower range tolerated by *E. coli* O157:H7. Under freezing conditions, non-pathogenic *E. coli* are reduced by 10-fold at −25.5°C over 38 weeks but little or no change in population numbers was noted for *E. coli* O157:H7 in ground beef over 9 months at −20°C (Doyle and Schoeni, 1984). In common with other vegetative bacteria the conditions of freezing and thawing (rates, method, temperature achieved, etc.) affect the degree of injury and death of cells of *E. coli*, but it is clear that survival of freezing processes can and does occur. Reliance on freezing to 'clean up' a contaminated food is therefore *not* a reliable practice.

Although *E. coli* will not grow under the normal refrigeration temperature conditions commonly used in the food industry, i.e. 3–7°C, pathogenic *E. coli* can survive well for several weeks with only a 0.5–1.5 log cycle reduction in populations noted over 1 to 5 weeks (International Commission on Microbiological Specifications for Foods, 1996a).

Pathogenic *E. coli* are not particularly heat resistant and in common with other Gram-negative pathogens thermal inactivation/heat resistance is affected by the other factors prevailing, such as pH (increases sensitivity

Table 3.1 Limits of temperature for the growth of pathogenic *E. coli* (adapted from International Commission on Microbiological Specifications for Foods, 1996a)

	Minimum (°C)	Optimum (°C)	Maximum (°C)
E. coli (all types)	7–8	35–40	44–46
VTEC O157:H7	8	37	44–45

Table 3.2 Studies of temperature effects on *E. coli* O157:H7

Substrate	Temperature (°C)	*D* value (min)	*z* value (°C)	Reference
Ground beef, 17–20% fat	60	0.75	4.1	Doyle and Schoeni, 1984
	62.8	0.4		
	64.3	0.16		
Ground beef	60			Ahmed *et al.*, 1995
7% fat		0.45	4.78	
10% fat		0.46	4.44	
20% fat		0.47	4.35	
Ground beef	63	0.43	5.6	Orta-Ramirez *et al.*, 1997
	68	0.12		
90% lean ground beef	55	21.13	6.0	Juneja *et al.*, 1997
	57.5	4.95 (2.7–7.83)		
	60	3.17 (0.61–4.54)		
	62.5	0.93 (0.39–1.37)		
	65	0.39 (0.16–1.45)		
Chicken	60			Ahmed *et al.*, 1995
3% fat		0.38	4.48	
11% fat		0.55	4.38	
90% lean ground chicken	55	11.83		Juneja *et al.*, 1997
	57.5	3.79		
	60	1.63		
	62.5	0.82 (0.48–1.31)		
	65	0.36 (0.21–0.52)		

to heat), water activity (a_w) and humectant used, fat content (can decrease apparent sensitivity to heat) and preconditioning of the organism, e.g. previous exposure to stress conditions (prior exposure to mild heat, i.e. heat shock) and/or anaerobic growth conditions (can decrease apparent sensitivity to heat).

Table 3.2 indicates the results from a variety of studies of the thermal inactivation of *E. coli* O157:H7. Other work has demonstrated that the organism shows greater survival following grilling of frozen ground beef patties than in patties held at higher temperature for a few hours prior to grilling, e.g. 21 or 30°C for 4 h, possibly due to the slower rate of heating in the frozen product or physiological changes in the cells of *E. coli* during freezing (Jackson *et al.*, 1996). In order to achieve at least a 10^6-fold (6*D* value) reduction of VTEC in beefburgers, a heat process equivalent to 70°C for 2 min is advocated for use in the UK (Advisory Committee on the Microbiological Safety of Food, 1995) (Table 3.3).

Table 3.3 Equivalent temperature/time heat treatments (Anon, 1992a)

Temperature (°C)	Time
60	45 min
65	10 min
70	2 min
75	30 s
80	6 s

For the consistent achievement of heat processes that will reliably reduce populations of *E. coli*, due account must be taken of the density, thickness, composition and initial temperature of the food.

pH, WATER ACTIVITY AND OTHER FACTORS

In addition to the application of good hygienic practices and well-controlled temperature regimes (heat or chill), other physico-chemical factors present in some specific food types can contribute to the control of the growth of any *E. coli* that may remain present in the food.

Table 3.4 summarizes some growth-limiting parameters for *E. coli*. For products in which the pH becomes non-optimal (usually acidic) either as a result of the manufacturing process, such as for cheese or fermented meats where lactic acid is produced by the metabolic activity of the starter cultures used, or by the direct addition of and mixing with an acidic component, e.g. oil and vinegar (acetic acid) dressing, the pH will contribute to the control of bacterial growth, including that of any *E. coli* that may be present. Organism survival depends on the type of acid present and other physico-chemical conditions prevailing.

Table 3.4 Growth-limiting parameters for pathogenic *E. coli* (adapted from International Commission on Microbiological Specifications for Foods, 1996a)

	Minimum	Maximum
pH	4.4*	9.0
a_w	0.95	–
Sodium chloride (Glass *et al.*, 1992)	Grows vigorously in 2.5% NaCl Grows slowly in 6.5% NaCl Does not grow in 8.5% NaCl	

* *E coli* O157:H7 is reported to be acid resistant, surviving at pH values below 4.4.

Table 3.5 Survival of *E. coli* O157:H7 in acidic foods

Suspect food	pH	Survival experiments	Reference
Apple cider unpasteurized	3.6–4.0	In ciders with no preservative added *E. coli* O157:H7 could be detected for up to 20 days when held at 8°C after initial inoculation with approximately 10^5 cfu/ml. When held at 25°C, survivors were detected at 2 and 3 days but not at 6 days after inoculation	Besser *et al.*, 1993 Zhao *et al.*, 1993
Live yoghurt, locally produced from pasteurized full-fat milk	4.5–4.6	*E. coli* O157:H7 (low inoculum, 10^3 cfu/ml, and high inoculum, 10^7 cfu/ml) was added at the same time as the starter cultures for making traditional and 'bifido' yoghurts. At both inoculum levels populations reduced to approximately 10% of the initial level after 7 days stored at 4°C. The final product pH was 4.5 in the traditional yoghurt and 4.6 in the 'bifido' yoghurt	Morgan *et al.*, 1993 Massa *et al.*, 1997
Mayonnaise and mayonnaise-based dressings and sauces	3.65–4.44	Three different strains of *E. coli* O157:H7 inoculated into mayonnaise at levels of approximately 10^7 cfu/g rapidly died off when stored at 25°C but cells were still detectable in mayonnaise up to 35 days when held at 7°C	Weagent *et al.*, 1994

Pathogenic *E. coli* will not grow in fermented cheeses at pH ≤5.4 in which lactic acid is present; *E. coli* O157:H7 will grow in a broth medium adjusted to pH 4.5 with hydrochloric acid but not when adjusted to the same pH with lactic acid (International Commission on Microbiological Specifications for Foods, 1996a). In some recent outbreaks of foodborne infection due to *E. coli* O157:H7 the organism has demonstrated significant acid tolerance (Table 3.5).

When organic acids (acetic, lactic, citric, propionic, etc.) are used as preservatives in food it is important to ensure that the correct concentration of undissociated acid (which is responsible for the antimicrobial activity) is available for bacterial growth inhibition. The proportion of undissociated acid present varies with pH (Table 3.6) so this must be taken into account when determining the amount of total acid required at a specific pH to give a particular concentration of undissociated acid. At neutral pH most organic acids will have a limited effect on the growth of *E. coli*.

Heat treatments applied to products with a sub-optimal pH for *E. coli* and/or containing an organic acid as a preservative are expected to be more effective than the same heat treatment applied to a product at the optimum pH for the organism.

The water activity of a food product is lowered by the addition of sodium chloride, sugars and/or other solutes. The higher the concentration of the solute, the lower the water activity of the product (Table 3.7). For products in which the water activity approaches or is lower than the minimum for growth of *E. coli*, growth of the organism will be prevented or minimized provided the water activity does not increase during the life of the product, e.g. by mixing the product with other foods of higher water activity or by allowing condensation to affect the product.

Table 3.6 Examples of the percentage of total undissociated organic acid present at different pH values (adapted from International Commission on Microbiological Specifications for Foods, 1980a)

Organic acid	pH value			
	4	5	6	7
Acetic acid	84.5	34.9	5.1	0.54
Citric acid	18.9	0.41	0.006	<0.001
Lactic acid	39.2	6.05	0.64	0.064

Table 3.7 Water activity at various concentrations of sodium chloride or sucrose (adapted from International Commission on Microbiological Specifications for Foods, 1980a)

Water activity (a_w)	Sodium chloride (%, w/w)	Sucrose (%, w/w, °Brix)
1.000	0	0
0.99	1.74	15.45
0.98	3.43	26.07
0.94	9.38	48.22
0.90	14.18	58.54
0.86	18.18	65.63

Recent outbreaks of infection with *E. coli* O157:H7 associated with dry fermented meat products of relatively low water activity (approximately 0.9) serve to underline the fact that this organism can survive the fermentation and drying processes used to produce such products. Thus it is important to use raw materials of high initial microbiological quality and ensure that processes are consistently well controlled to prevent the growth of contaminating organisms, including any *E. coli* that may be present.

In 1994 an outbreak of 23 cases of infection due to *E. coli* O157:H7 in Washington state and California was attributed to presliced dry fermented salami (Alexander *et al.*, 1995). Samples of implicated salami were shown to have an internal pH range of 4.9–5.0, a salt content range of 3.7–3.9% and protein–moisture ratios in the range 1.8 to 1.86; the estimated number of *E. coli* O157:H7 organisms consumed by each of four patients studied was less than 50 (Tilden *et al.*, 1996). The investigation carried out and reported by Tilden *et al.* (1996) yielded no evidence of post-processing contamination and the outbreak was likely to have been due to raw material contamination as it had already been demonstrated that the organism can survive the relatively harsh manufacturing conditions of such products, which takes place over several weeks (Glass *et al.*, 1992; Hinkens *et al.*, 1996). Table 3.8 summarizes the key product characteristics of the dry fermented meat products in which *E. coli* O157:H7 survived as reported by these workers. The results of these studies demonstrate that combining sub-optimal physico-chemical conditions such as pH, temperature and water activity has a greater effect than any of the individual factors used at the same level.

A considerable amount of work is still being carried out around the world to gain further knowledge and understanding of the effects of current and

Table 3.8 Dry fermented meat product characteristics at the end of fermentation and drying processes in which *E. coli* O157:H7 survived.

	Glass *et al.*, 1992[a]	Hinkens *et al.*, 1996[b]
pH	4.5	4.74 ± 0.09
Moisture content (%)	37	27.47 ± 1.97
Fat content (%)	31	46.6 ± 1.44
Sodium chloride concentration (%)	4.9	4.4 ± 0.29
Protein content (%)	21	20.33 ± 0.73
Moisture/protein ratio	1.8:1	1.35 ± 0.14:1
Water activity	?	0.87 ± 0.02

[a] Values are an average of three determinations for each of two trials at the end of drying for 18 to 21 days.
[b] After 21 days drying.

novel food industry practices and processes, including the effects of individual and combined physical and chemical systems on the survival and growth of VTECs in foods.

It will always remain essential that food producers and processors maintain a thorough knowledge and understanding of the potential sources of and particular requirements for control of VTEC in their specific areas of activity.

4

INDUSTRY FOCUS: CONTROL OF *E. COLI*

INTRODUCTION

The evidence from recent outbreaks relating to the low infective dose required to cause illness associated with Vero cytotoxin-producing *E. coli* clearly demonstrates the need for the food industry to employ measures to prevent the organism from being present in foods at the point of consumption. Many products can present a potential risk of causing outbreaks of *E. coli* food poisoning but, in most outbreaks, a failure in the control systems can be identified as a major contributory factor to the outbreaks. In such cases, hazard analysis and implementation of controls at the critical points identified could have prevented the outbreaks, provided the control systems were operated consistently correctly. It is strongly recommended that all persons involved in the primary production, processing and sale of food adopt a hazard analysis approach considering all relevant pathogens, including Vero cytotoxin-producing *E. coli*. Indeed, the requirement to operate such an approach in food businesses is embodied in European and national laws (Anon., 1993b; Anon., 1995d).

To help focus attention on the products representing the greatest concern in relation to Vero cytotoxin-producing *E. coli* and the areas requiring greatest management control, a series of questions can be applied to each food process and product. Processes and products can be reviewed against these key questions to identify the level of concern that Vero cytotoxin-producing *E. coli* may represent (Table 4.1). As a guide to answering these questions, some familiar products in different commodity groups are given as examples (Table 4.2). After answering each of the questions in Table 4.1 the product can be assessed against the profiles given in Table 4.3 to determine the level of concern that may be associated with the product. Having done this, the key process areas

Table 4.1 How much of a concern does your product represent?

Question	Yes	No
Is *E. coli* expected to be present in the raw material?		
Is the raw material of bovine origin or exposed to contamination from bovine sources?		
Will the organism be destroyed or reduced to an acceptable level by any of the processing stages?		
Will the product be exposed to any post-process contamination?		
Can the normal process conditions allow the organism to grow?		
Will the product be subjected to a process by the customer that will destroy *E. coli*?		

requiring greatest attention for control of the hazard can be determined (Table 4.4).

Every process and product will differ from those presented in the tables therefore the tables should be used for guidance purposes only. Complete understanding of the hazards and controls can only be gained by applying a full hazard analysis. In addition, it is important to note that even processes and products that are rated as being of very low concern in relation to Vero cytotoxin-producing *E. coli* may still be capable of causing outbreaks if the controls inherent in the normal manufacture of these products are not applied correctly. In fact, significant hazards to food safety are presented by complacent management teams who believe that their product is safe because of historical precedence or by a food production team lacking the necessary skills and training in safe food manufacture. Food products are generally made safe or unsafe to eat by human intervention.

The highest concern products with regard to Vero cytotoxin-producing *E. coli* are those where the organism may be present, even in low numbers, in the raw material, where no process exists to reduce or eliminate it and where it may survive or even grow during the process or in the finished product, which is consumed without any further processing. Products such as raw-milk mould-ripened soft cheeses fall within this category and it is surprising that such products have so far rarely been implicated in outbreaks of illness caused by Vero cytotoxin-producing *E. coli*.

Table 4.2 Examples of the key process stages where *E. coli* O157 or VTEC may represent a hazard in different foods*

Product	Product examples	Raw material contamination	Raw material of bovine origin	Reduction process	Destruction process	Post-process contamination	Process allows growth	Consumer cidal process
Dairy products								
Raw-milk ripened soft cheese	Raw-milk Brie, Camembert	Yes	Yes	No	No	Yes	Yes	No
Raw-milk hard cheese	Raw-milk cheddar, parmesan	Yes	Yes	Yes	No	Yes	No	No
Pasteurized-milk-ripened soft cheese	Brie, Camembert	Yes	Yes	Yes	Yes	Yes	Yes	No
Pasteurized milk hard cheese	Edam, Cheddar, Cheshire	Yes	Yes	Yes	Yes	Yes	No	No
Pasteurized milk fermented products	Yoghurt, fromage frais, cottage cheese	Yes	Yes	Yes	Yes	No	No	No
Meat products								
Raw meat and poultry	Pork, lamb, chicken	Yes	No	No	No	Yes	No	Yes
Raw meat	Beef/comminuted beef products	Yes	Yes	No	No	Yes	No	Yes
Fermented and dry-cured meat	Salami, Parma ham	Yes	Yes/No	Yes	No	Yes	No	No

Table 4.2 Continued

Product	Product examples	Raw material contamination	Raw material of bovine origin	Reduction process	Destruction process	Post-process contamination	Process allows growth	Consumer cidal process
Cooked meat (hermetically sealed)	Whole hams, chub pâté	Yes	Yes/No	Yes	Yes	No	No	No
Cooked sliced meat	Ham, chicken, frankfurters, pâté	Yes	Yes/No	Yes	Yes	Yes	No	No
Fish and shellfish								
Raw fish/shellfish	Cod, mussels	Yes	No	No	No	Yes	No	Yes
Raw fish/shellfish, consumed raw	Oysters, sushi	Yes	No	No	No	Yes	No	No
Cold-smoked fish, consumed raw	Smoked salmon and smoked trout	Yes	No	No	No	Yes	No	No
Cooked fish/shellfish/fish pâté	Prawns, crab	Yes	No	Yes	Yes	Yes	No	No
Salad and vegetables								
Raw vegetables	Potatoes, broccoli, beans, peas	Yes	No	No	No	Yes	No	Yes
Raw salads	Lettuce, spring onions, celery	Yes	No	No	No	Yes	No	No

Table 4.2 Continued

Product	Product examples	Raw material contamination	Raw material of bovine origin	Reduction process	Destruction process	Post-process contamination	Process allows growth	Consumer cidal process
Prepared salads	Salad mix	Yes	No	Yes	No	Yes	No	No
Sprouted vegetables	Beansprouts, alfalfa sprouts, salad cress	Yes	No	Yes	No	Yes	Yes	No

Raw material contamination: Is *E. coli*, *E. coli* O157 or VTEC expected to be present in the raw material?
Raw material of bovine origin: Is the raw material of bovine origin or exposed to contamination from bovine sources?
Reduction or destruction process: Will the organism be reduced to an acceptable level or destroyed by any of the processing stages?
Post-process contamination: Will the product be exposed to any post-process contamination?
Process allows growth: Are the normal processing conditions suitable for the growth of *E. coli* O157 or VTEC, if present?
Consumer cidal process: Will the product be subjected to a process by the customer that will destroy *E. coli* O157 or VTEC?

* Information given is for guidance only and may not be appropriate for individual circumstances. It is recommended that proper hazard analysis is carried out for every process and product to identify where controls must be implemented to minimize the hazard from *E. coli* O157 or VTEC.

Table 4.3 Categories of concern*

Level of concern	Product examples	Raw material contamination origin	Raw material of bovine	Reduction process	Destruction process	Post-process contamination growth	Process allows process	Consumer cidal
Category 1: Highest	Raw-milk ripened soft cheese	Yes	Yes	No	No	Yes	Yes	No
	Sprouted vegetables, beansprouts, alfalfa	Yes	No	Yes	No	Yes	Yes	No
Category 2: High	Salami, dry-cured ham	Yes	Yes/No	Yes	No	Yes	No	No
	Raw-milk hard cheese	Yes	Yes	Yes	No	Yes	No	No
	Prepared salads	Yes	No	Yes	No	Yes	No	No
	Oysters, sushi, smoked salmon	Yes (limited)	No	No	No	Yes	No	No
Category 3: Medium	Raw beef/comminuted beef products, burgers	Yes	Yes	No	No	Yes	No	Yes
Category 4: Medium	Fresh-pressed juice	Yes	No	Yes	No	No	No	No
	Pasteurized milk ripened cheese, Brie	Yes	Yes	Yes	Yes	Yes	Yes	No
Category 5: Low	Cooked sliced beef, beef pâté	Yes	Yes	Yes	Yes	Yes	No	No
	Pasteurized milk hard cheese cheddar, edam	Yes	Yes	Yes	Yes	Yes	No	No

Table 4.3 Continued

Level of concern	Product examples	Raw material contamination	Raw material of bovine origin	Reduction process	Destruction process	Post-process contamination	Process allows growth	Consumer cidal process
Category 5: Low	Cooked sliced chicken, ham, pork, pâté, fish, crab	Yes	No	Yes	Yes	Yes	No	No
Category 6: Low	Chub pâté, products cooked in pack	Yes	Yes/No	Yes	Yes	No	No	No

High concern: Where *E. coli* O157 or VTEC could be present due to raw material contamination or as a post-process contaminant and where the process allows survival and/or growth and the product is ready to eat.
Medium concern: Where *E. coli* O157 or VTEC may be present in the raw material or as a post-process contaminant and where the process usually achieves a reduction in the organism or a consumer cidal process is applied.
Low concern: Where *E coli* O157 or VTEC may be present in the raw material but the process applied destroys the organism and there is limited opportunity to recontaminate the product.

* Information given is for guidance only and may not be appropriate for individual circumstances. It is recommended that proper hazard analysis is carried out for every process and product to identify where controls must be implemented to minimize the hazard from *E. coli* O157 or VTEC.

Table 4.4 Process stages where control of *E. coli* O157 or VTEC is critical (based on the categories of concern)*

Level of concern	Product examples	Raw material control	Reduction process	Destruction process	Post-process contamination	Process conditions	Consumer issues
Category 1: Highest	Raw-milk ripened soft cheese	Yes			Yes		Yes
	Sprouted vegetables, beansprouts, alfalfa	Yes	Yes		Yes	Yes	
Category 2: High	Salami, dry-cured ham	Yes	Yes		Yes	Yes	
	Raw-milk hard cheese	Yes	Yes		Yes	Yes	
	Prepared salads	Yes	Yes		Yes		
	Oysters, sushi, smoked salmon	Yes	Yes				Yes
Category 3: Medium	Raw beef/comminuted beef products, burgers	Yes			Yes		Yes
Category 4: Medium	Fresh-pressed juice	Yes	Yes			Yes	
	Pasteurized milk ripened cheese, Brie	Yes		Yes	Yes		
Category 5: Low	Cooked sliced beef, beef pâté	Yes		Yes	Yes		
	Pasteurized milk hard cheese cheddar, edam	Yes		Yes	Yes		

Table 4.4 Continued

Level of concern	Product examples	Raw material control	Reduction process	Destruction process	Post-process contamination	Process conditions	Consumer issues
Category 5: Low	Cooked sliced chicken, ham, pork, pâté, fish, crab			Yes	Yes		
Category 6: Low	Chub pâté, products cooked in pack			Yes			

* Information given is for guidance only and may not be appropriate for individual circumstances. It is recommended that proper hazard analysis is carried out for every process and product to identify where controls must be implemented to minimize the hazard from *E. coli* O157 or VTEC.

RAW FERMENTED AND DRY-CURED MEAT PRODUCTS

Raw fermented meat products have been associated with a number of food poisoning outbreaks involving VTEC (Alexander *et al.*, 1995; Cameron *et al.*, 1995a). Raw fermented meat and raw dry-cured meat products represent high risk product groups in relation to enteric pathogens because of the nature of the raw materials and manufacturing processes employed. Fermented meats, traditionally called salamis, include products such as German salami, peppered salami and Danish salami. Raw dried meats, on the other hand, often command a premium position in the market with products like Parma ham having a world-wide distribution. These products are made using long-established traditional processes and the countries most commonly associated with their manufacture are those in continental Europe, including Italy, France and Germany.

Fermented meats and raw dried meats differ markedly in their production processes and, as a consequence, the risk associated with the survival of contaminating pathogens is likely to differ, with salamis considered to be of greatest risk.

Description of process

Fermented meat products are usually manufactured by comminuting meats and fat and mixing them with herbs, spices, salt, sugar and the preservative sodium nitrite during a bowl chopping operation (Figure 4.1). Many salamis also have a bacterial starter culture added to the meat mix although some traditional processes rely on the development of natural lactic microflora as part of their process. Meat used in the manufacture of salami is predominantly pork although many salamis are manufactured with pork and beef and occasionally include other species, such as turkey. The meat mix, or batter as it is often called, is forced into a casing, which may be made of synthetic material or may be derived from the intestines of animals, usually sheep. The salami sausage is tied with string and this is used to suspend it on a rack with other salamis. Each batch is placed in a fermentation room where the conditions of temperature and relative humidity are usually carefully controlled. Fermentation temperatures may vary depending on the type of salami, with the greatest differences usually between salamis manufactured in Europe and the USA. In the USA, salami is usually fermented at temperatures up to and in excess of 30°C whereas in Europe temperatures are usually much lower, between 20 and 30°C. Such differences in temperature will usually result in different rates of fermentation with the American salamis achieving a faster reduction in pH and acidity development. During fermentation, the pH of the product

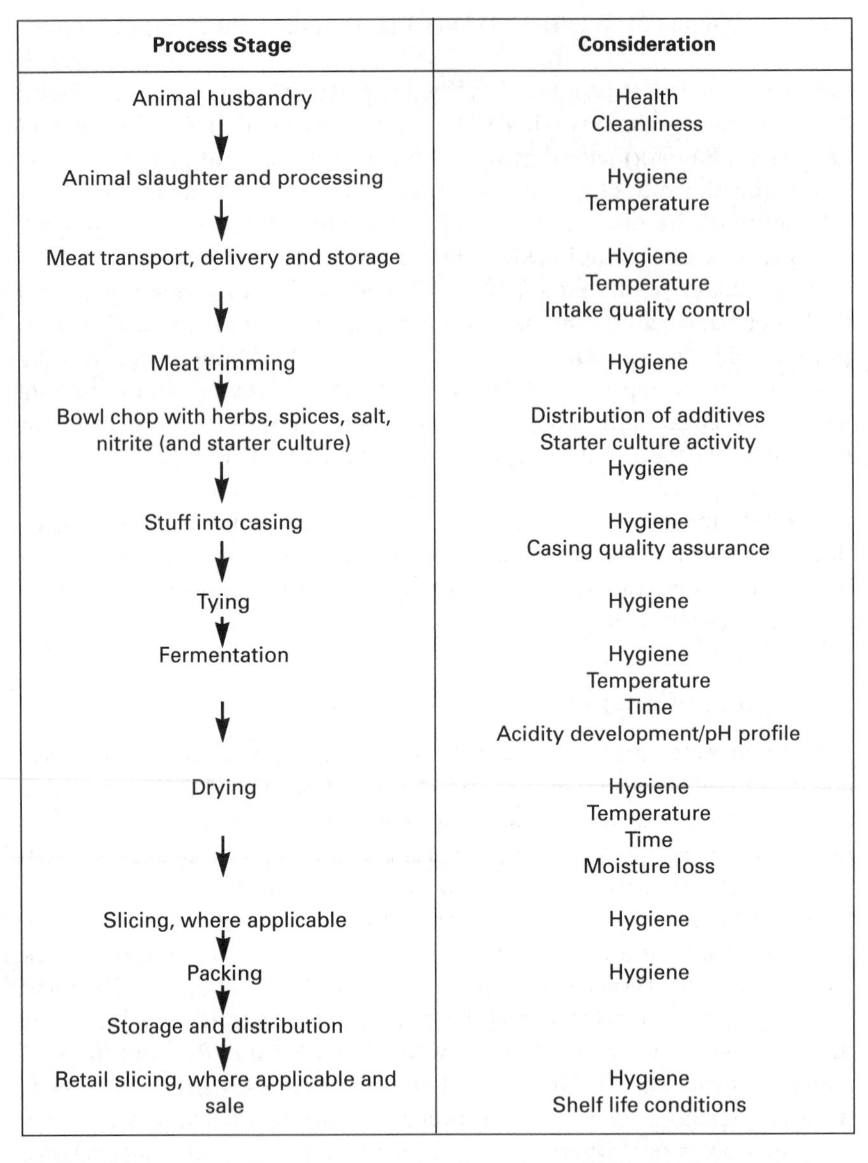

Process Stage	Consideration
Animal husbandry	Health Cleanliness
Animal slaughter and processing	Hygiene Temperature
Meat transport, delivery and storage	Hygiene Temperature Intake quality control
Meat trimming	Hygiene
Bowl chop with herbs, spices, salt, nitrite (and starter culture)	Distribution of additives Starter culture activity Hygiene
Stuff into casing	Hygiene Casing quality assurance
Tying	Hygiene
Fermentation	Hygiene Temperature Time Acidity development/pH profile
Drying	Hygiene Temperature Time Moisture loss
Slicing, where applicable	Hygiene
Packing	Hygiene
Storage and distribution	
Retail slicing, where applicable and sale	Hygiene Shelf life conditions

Figure 4.1 Process flow diagram and technical considerations for a typical raw fermented meat.

reduces to values usually approaching or slightly below pH 5.0. The inherent safety of the product is highly dependent on the fermentation stage to reduce pH, together with the concomitant production of organic acids by the lactic microflora. After several days of fermentation the salami is dried at lower temperatures (<15°C) for several weeks during which time the moisture content of the product and its water activity markedly reduce.

Many of the traditional European salamis also undergo mould ripening during the drying process, the mould developing on the external surface of the salami and imparting a characteristic flavour. The growth of mould is usually accompanied by the elevation of pH, particularly at the surface of the product, and, depending on the degree of ripening, this can result in a pH increase from pH <5.0 to pH 6-7. The pH of mould-ripened salami may vary considerably across the salami stick with the lowest pH being at the furthest point away from the mould growth, i.e. the centre, and the highest pH being on the surface. Mould may be removed from the external surface of the salami by brushing prior to packing and sale. Salamis usually have a final pH of 4.5-5.0 although, as described, mould-ripened varieties may be of higher pH, between 5.5 and 7. Depending on the extent of drying, salamis have an aqueous salt content of 5 to 10% and a water activity of <0.94, although this may be as low as 0.85 or less. Again, the water activity can vary across the salami and this is most affected by the thickness of the salami itself, with more even moisture loss occurring in thin salamis than in thick ones. In contrast to pH, the water activity is usually lowest at the surface and highest in the centre. Some salamis may be smoked prior to the drying stage by holding in smoking rooms filled with natural wood smoke at temperatures below 15°C. Salami may be sold as a bulk product for cutting or slicing on the retail delicatessen counter or it may be sliced at the factory and sold as a prepack. The low pH, high aqueous salt content and low water activity of salami usually allow long product shelf lives of several months as few micro-organisms are capable of growth on a finished salami product. For retail purposes the products are often sold from a chilled cabinet but many salamis could actually be considered to be ambient stable products.

Raw dry-cured meats differ significantly from salami in the nature of their production. These products are usually manufactured from whole anatomical pieces of meat. The raw meat is usually pork although some products are made with beef. The raw meat is trimmed to achieve the desired fat content and then salted prior to storage under refrigerated conditions (<5°C) for several weeks (Figure 4.2). During this time the meat is turned frequently, with fresh curing salts being applied to ensure even distribution of salt throughout the raw meat. The conditions of refrigerated storage and high salt content select a dominant microflora of lactic acid bacteria. Following the refrigerated storage the salted raw meats are dried at temperatures of <15°C for several months. During this stage a mild fermentation occurs, leading to a slight decrease in the pH. Some products may be smoked prior to drying. After drying the product usually achieves a pH between 5 and 6 and a water activity of 0.85-0.90. Like salamis, raw dried meat products may be sold bulk or sliced for

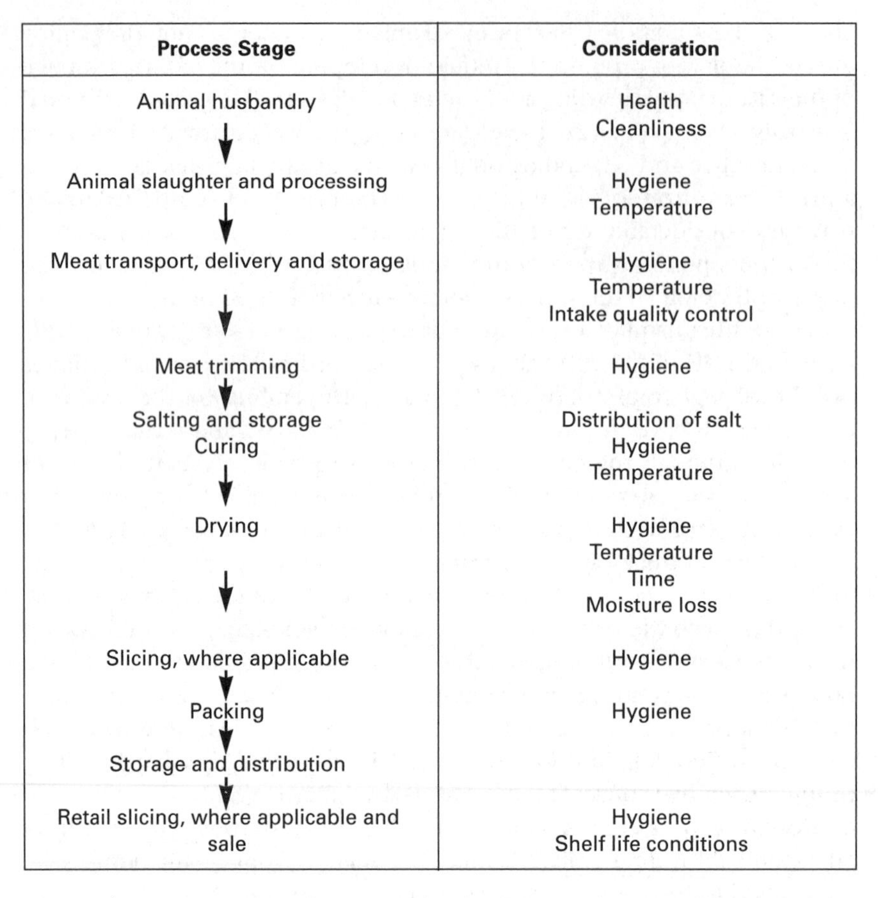

Process Stage	Consideration
Animal husbandry	Health Cleanliness
Animal slaughter and processing	Hygiene Temperature
Meat transport, delivery and storage	Hygiene Temperature Intake quality control
Meat trimming	Hygiene
Salting and storage Curing	Distribution of salt Hygiene Temperature
Drying	Hygiene Temperature Time Moisture loss
Slicing, where applicable	Hygiene
Packing	Hygiene
Storage and distribution	
Retail slicing, where applicable and sale	Hygiene Shelf life conditions

Figure 4.2 Process flow diagram and technical considerations for a typical raw dry-cured meat.

prepack sale. Raw dried meats are considered to represent a lesser hazard in relation to enteric pathogen survival as the contaminants are likely to be restricted predominantly to the external surface of the meat, which coincides with the presence of high concentrations of inhibitory factors such as salt and the rapid development of competitive lactic microflora. The adverse effect of these factors on the contaminating pathogens is likely to confer a greater degree of security than that existing in a salami process. Subsequent sections will focus primarily on issues relating to fermented meats.

Raw material issues and control

The major raw material concern in relation to salami and raw dried meat manufacture is the raw meat itself. VTEC can be present in almost all species of raw meat (Table 1.6). As many of the outbreaks associated with *E. coli* O157 have been associated with products of bovine origin, the nature of the meat itself directly affects the inherent risk associated with the different types of product. The presence of beef may be considered to represent a greater hazard to the process than other species of meat and as small quantities of beef are often included in many salami products, the exclusion of beef may reduce the risk associated with those products, often without having much effect on organoleptic quality. However, it is clear that *E. coli*, VTEC and *E. coli* O157 may all occur on other species of meat at differing frequencies and therefore the hazard cannot be dismissed in relation to other meats. For example, pork, which is the predominant meat used in the production of salami, was found to be contaminated with VTEC in 4% of samples tested (Wray *et al.*, 1993). In addition, it important to note that data on the incidence of VTEC are not comprehensive and even if the incidence is low in any particular species of meat at the moment this by no means guarantees that it will remain so in future years. In addition, if the raw material received by the salami manufacturer is not a carcass or primal joint but a further processed cut of meat it is possible that more than one species may have been processed by the raw material supplier. In this circumstance the opportunities for VTEC incidence may be increased due to cross-contamination of the pork from production lines used in common with beef. Whether beef, pork or any other species of meat is used, limiting raw material contamination is critical to the safety of the salami process and effective supplier quality assurance programmes are essential to reduce the risk associated with the hazard of *E. coli*. Standards of hygiene at supplying slaughterhouses should be monitored by regular inspection and auditing programmes conducted by the salami or raw dried meat manufacturer. Such raw material supplier auditing needs to stress the importance of minimizing contamination with enteric pathogens. Suppliers must have a clear understanding that the raw material is to be used for products not subjected to any cooking stage and a partnership approach between the manufacturer and their raw meat supplier can have significant dividends in the provision of a safer raw material. As an additional part of any supplier quality assurance plan it is important to incorporate regular monitoring of the raw meat on arrival. Monitoring for indicators of hygienic processing such as *E. coli* may provide valuable information for trend analysis to provide early warning of any decreasing standards of hygiene at the supply site.

The primary stage in the manufacture of salami and raw dried meat usually involves processing primal cuts of raw meat and, where appropriate, trimming excess fat to achieve the desired fat content consistent with achieving the appropriate visual and chemical condition of the subsequent product. Any processing of the raw meat will inevitably create opportunity for further cross-contamination from contaminated meat to uncontaminated batches because of the use of common conveyors, knives or other butchery utensils, and, of course, personnel themselves. High standards of hygiene are critical to ensure cross-contamination is minimized and this should include regular cleaning and sanitization of knife blades and other utensils. In butchery environments it is inevitable that pathogens will be present in the environment from contaminated products and it is imperative to prevent build up of contaminants in all processing areas and on equipment. Attention to effective cleaning of product contact surfaces and the environment with appropriate cleaning and sanitization programmes will significantly reduce the extent of cross-contamination with pathogens.

It is easy for operatives handling raw meat to consider that because of the visual appearance of the raw material it may already be contaminated with 'harmful bacteria', i.e. pathogens, and therefore high standards of hygiene are of little importance. Such complacency could lead to the build-up of pathogens like VTEC in processing areas, creating focal points of contamination and the potential for extensive cross-contamination to clean raw material batches. Any increase in levels or incidence of VTEC in the raw meat will ultimately compromise the safety of the product. The emphasis on high standards of hygiene at this early stage is critical to limiting the extent of raw material contamination. However, even with the highest standards of hygiene it is anticipated that salami and raw dried meat processes will be challenged from time to time with enteric pathogens from contaminated raw materials; clearly this is likely to include VTEC, including *E. coli* O157.

While the raw meat must represent the focus for control of enteric pathogens, it is important to remember that products such as salamis also include a variety of other potential hazards. Herbs and spices are frequently reported to be contaminated with pathogens such as *Salmonella*, although this is dependent on their origin and treatment received. The presence of such pathogens is likely to have arisen due to faecal contamination during growing, harvesting and drying, and opportunities may also exist for VTEC to be present from such sources. The incidence of *E. coli* O157 in such raw materials has not yet been extensively studied but the hazard should be considered as part of the hazard analysis conducted by the producers of these materials and appropriate controls implemented into supplier quality assurance schemes for these raw materials. Like raw meat, such schemes may include monitoring of the raw materials for specific pathogens or for indicator microorganisms. Alternatively, certificates of analysis, including relevant microbiological information, may be requested from the raw material supplier. Wherever possible, herbs and spices should be sourced from manufacturers employing a decontamination method such as steam pasteurization. A further critical raw material in the salami process is the casing. Synthetic casings are unlikely to introduce microbial pathogens but traditional processes also involve the use of natural casings from animals, including sheep and cows, and the use of such casings has obvious associated contamination risks from enteric pathogens. Natural casings are exposed to harsh treatments during processing, including washing, salting and drying. Prior to use in the salami process the casing is rehydrated. The effectiveness of the treatments in reducing enteric pathogens such as *E. coli* O157 must be considered as an integral part of the hazard analysis for the supply of natural casings.

Process issues and control

In a salami process, raw meat is usually mixed with herbs and spices at the bowl chopping stage and, given the fact that the presence of sodium nitrite and salt is likely to be an important contributory factor in controlling the growth and survival of enteric pathogens, it is essential that the ingoing levels of these substances are closely monitored and that an even distribution is achieved throughout the meat mix. Products to which starter cultures are added rely on the activity of the organisms used therefore the inoculum levels and starter culture activity are important controlling factors in the manufacturing process. Where starter cultures are not added, the fermentation is reliant on the natural microflora present. Processes are usually monitored by measuring changes in the chemical indicators of microbial activity, such as pH and titratable acidity, rather than levels of the organisms themselves.

In challenge test studies of fermented meats manufactured with or without added starter cultures there was no significant difference observed in the survival of *E. coli* O157 at the end of the process (Glass *et al.*, 1992).

In terms of the process, it is clear that at the bowl chopping stage there is an opportunity for spreading any contamination to each batch of meat mix. It is therefore important to maintain the equipment in a hygienic condition. Bowl choppers are difficult to clean effectively, usually harbouring contamination in the recesses of the blade housing, around seals and in the bowl itself. Close attention to thorough cleaning and disinfection needs to be given and this should involve regular dismantling of the equipment, including the blade. Monitoring post-cleaning hygiene of the equipment, using bacterial indicators of hygiene such as Enterobac-teriaceae or even rapid ATP bioluminescence hygiene assays, can give useful information regarding cleaning efficacy.

The primary reason why traditional salamis represent such a major food safety hazard in relation to enteric pathogens like VTEC is because of the distribution of such contaminants throughout the raw meat mix. A further opportunity for cross-contamination to salamis occurs during the casing process where contamination could be introduced or spread by operatives from environmental sources due to handling of casings during stuffing. Although less significant than the likelihood of contamination from the raw meat itself, such areas should not be neglected in hazard analysis studies to ensure appropriate hygiene and cleaning regimes are in place to contain any hazards.

One of the most critical stages in the manufacture of salami is the fermentation process. The cased raw meat sausage is tied to a frame and placed in a fermentation chamber of controlled humidity and temperature. The design of the chamber and frames allows free airflow between the products and effective moisture loss during the process. Depending on the process, the product is fermented for two to five days. In the USA processes involve higher temperatures for shorter times whereas European processes usually operate at lower temperatures for extended periods. The fermentation process in both cases is designed to achieve the same effect, i.e. to encourage the growth of lactic acid bacterial microflora with the production of organic acids and a decrease in pH. Although processes using added starter cultures are more controllable, it is important to ensure that the fermentative microflora is active, as demonstrated by the development of acidity or a decrease in the pH value of the product. Without a sufficient growth rate of this competitive microflora significant opportunities exist for the growth of contaminating pathogens, including *E. coli* O157. Some components of the raw meat mix itself may have a slight inhibitory effect on the growth of *E. coli* but it is important to note that the mix will usually have a very high water activity (>0.97) with aqueous salt contents of 2 to 5%, little acidity and pH values between 5.8 and 6.5. Although sodium nitrite is added to the meat mix, this is unlikely to have a significant inhibitory effect on the growth of *E. coli* O157 during the fermentation or subsequent processes. At best, the fermentation stage is likely to prevent significant growth of contaminating pathogens, but without the active growth of lactic acid bacteria and production of further inhibitory compounds it is certainly possible to predict that pathogens like *E. coli* O157 may be capable of growth under the prevailing conditions. Gibson and Roberts (1986) demonstrated that enteropathogenic *E. coli* could grow in laboratory media with up to 200 ppm nitrite, 6% sodium chloride and at a pH of 5.6. Effective growth of the lactic microflora will be dependent on the culture characteristics as well as the times and temperatures employed during fermentation, therefore it is critical that salami manufacturers clearly establish and control fermentation profiles in three key areas: the initial meat mix temperature, the humidity and the environmental temperature. Any changes made to these conditions need to take account of the impact such changes may have on the growth and acidity development of the lactic microflora. During fermentation the pH of a typical salami may decrease from commonly pH 6–6.5 to below pH 5.0. This naturally varies, with some products remaining above pH 5.0 and some decreasing to pH 4.5. Although the acidity increase and pH decrease have an impact on the ultimate safety of the product, the levels achieved are usually dictated by the contribution they make to the final taste of the product and is therefore more guided by the desire for a certain flavour than the elimination of pathogenic contaminants.

Whilst the fermentation process is important primarily for the development of the microflora and associated acidity, it should not be overlooked that significant moisture loss occurs at the elevated temperatures involved and this in turn results in the concentration of antimicrobial factors such as acidity, salt and nitrite in the aqueous phase. In a typical fermentation process, the moisture content may be reduced by 10 to 15%. Following fermentation, salami products are usually moved to a second chamber where the temperature is reduced to below 15°C and the humidity is controlled. The product is then dried for extended periods to reduce the moisture content even further. It is usually during this extended drying stage that most destruction of contaminating enteric pathogens occurs. The effect of organic acids, reduced pH, salt content and nitrite, all concentrated into the reduced aqueous phase due to moisture loss, has the potential to reduce bacterial contaminant levels significantly. In challenge test studies conducted on salami-type products it is clear that enteric pathogens can be reduced in number during the salami process. Glass *et al.* (1992) investigated the survival of *E. coli* O157:H7 during the manufacture of a fermented meat product. The process involved a high temperature fermentation (35.6°C for 13–14 h) of a pork–beef batter mix that contained 156 ppm sodium nitrite and 3.5% salt. The post-fermentation pH was 4.8 and the sausages were then dried at 12.8°C for 18 to 21 days prior to vacuum packing and storage at 4°C for 8 weeks. *E. coli* O157 levels decreased only slightly during the fermentation stage at 35.6°C (0.32 log cfu per gram). The levels decreased by a further 0.6 log unit during the drying stage and by 1 log cfu/g during storage at 4°C for 8 weeks. In comparing a number of trials it was evident that longer drying conditions were an important factor in destroying more *E. coli* O157:H7. It was concluded that inactivation was primarily due to acidity development and drying.

Following an outbreak of *E. coli* O157 food poisoning linked to salami in the USA (Alexander *et al.*, 1995), the United States Department of Agriculture (USDA) Food Safety Inspection Service issued a notice to the industry requiring that fermented meat processes should achieve a 5 log cfu/g reduction of contaminating *E. coli* O157 or, failing this, the implementation of a quality control programme ensuring high quality raw material use together with a statistically significant sampling programme on raw material and finished product (Anon., 1996b). A Blue Ribbon Task Force on *E. coli* O157:H7 established by the National Cattlemen's Beef Association in the USA co-ordinated some excellent research conducted at the Food Research Institute of the University of Wisconsin on the survival of *E. coli* O157:H7 in dry fermented sausage (Anon., 1996b). They studied the effect of fermentation temperature, pH development, drying, salami diameter and pasteurization on *E. coli* O157:H7 inoculated as a five-strain cocktail into raw batter at $>10^7$ cfu per gram. The studies concluded that most processes trialed would result in a 2 log cfu/g reduction in *E. coli* O157:H7 but many could not achieve a full 5 log cfu/g reduction (Tables 4.5 and

Table 4.5 Profiles of salami products not achieving a 5 log reduction in *E. coli* O157 (adapted from Anon., 1996b)

Fermentation condition	pH after fermentation	Cooking stage	Drying or holding condition	Salami size*
21.1°C (70°F)	4.6	No	Dry or hold at 21.1°C (70°F) for 7 days and then dry	Small casing
32.2°C (90°F)	4.6	No	Hold at 32.2°C (90°F) for 7 days then dry	Large casing
32.2°C (90°F)	5.3	No	Hold at 32.2°C (90°F) for 7 days and then dry	Large casing
43.3°C (110°F)	4.6	No	Dry for 4 days	Small and large casing

* Small casing = 55 mm diameter; large casing = 105 mm diameter.

Table 4.6 Profiles of salami products achieving a 5 log reduction in *E. coli* O157 (adapted from Anon., 1996b)

Fermentation condition	pH after fermentation	Cooking stage	Drying or holding condition	Salami size*
32.2°C (90°F)	5.3	Yes	Dry for 7 or more days	Large casing
32.2°C (90°F)	4.6	No	Hold at 32.2°C (90°F) for 6 or more days	Small casing
32.2°C (90°F)	4.6	Yes	-	Small or large casing
43.3°C (110°F)	4.6	No	Hold at 43.3°C (110°F) for 4 or more days	Small or large casing

* Small casing: 55 mm diameter; cooking profile: 1 h at 37.7°C (100°F) and 6 h at 51.6°C (125°F). Large casing: 105 mm diameter; cooking profile: 1 h at 37.7°C (100°F), 1 h at 43.3°C (110°F), 1 h at 48.8°C (120°F) and 7 h at 51.6°C (125°F).

4.6). This is consistent with the findings of other challenge tests conducted on salami in recent years. The work identified clear factors that could be considered to increase or decrease the risk associated with these processes and these should be considered by all persons involved in the manufacture of these types of products (Table 4.7).

Other survival studies with *E. coli* O157:H7

Studies have been conducted to assess the survival of *E. coli* O157:H7 in beef jerky, a commonly consumed dried, ambient stable meat product in the USA. Beef jerky is manufactured by drying marinated beef strips to low levels of moisture, resulting in water activity of less than 0.9. Harrison and Harrison (1996) reported that during the initial drying stage of marinated beef at 60°C the population of *E. coli* O157:H7 decreased by over 3 log cfu/g within 3 h and over 5 log units within 10 h. During subsequent storage at ambient temperature at water activities of 0.75, 0.84 and 0.94 no survivors could be found during the 8 week study. Harrison *et al.*, (1997) also showed that reductions in *E. coli* O157:H7 were 1–2 log cfu/g greater in high salt (>1%) ground beef jerky than low salt (<0.1%) varieties.

Hinkens *et al.* (1996) investigated the survival of *E. coli* O157:H7 in a pepperoni manufacturing process and also assessed the impact of various heat processing stages after fermentation on the survival of these pathogens.

Table 4.7 Risk factors associated with the survival of *E. coli* O157:H7 in salami processes (adapted from Anon., 1996b)

Factor	Risk of survival	
	Higher	Lower
Beef as a main ingredient	Yes	
Indication of poor hygienic processing, i.e. high initial coliform count on meat	Yes	
Fermentation at low temperature	Yes	
Fermentation to a high pH	Yes	
Application of a heat destruction process		Yes
Drying to a low water activity		Yes

The pepperoni was fermented at 36°C until a pH of 5.0 (14 to 18 h) was achieved and then it was dried at 13°C for 15 to 21 days. Where a heat process was applied, this was conducted after the fermentation stage and included instantaneous heating at 63°C or heating at 53°C for 60 min.

In the standard pepperoni process (without heat processing), initial contamination levels of log 6.89 cfu per gram remained fairly constant during the fermentation stage with no appreciable decrease. Levels decreased by nearly 0.5 log cfu/g within 8 days of drying and by the end of the process (day 21) final levels reached log 5.69 cfu/g. In total, a decrease of less than 2 log cfu/g in *E. coli* O157:H7 was achieved by the standard process. Heating the product to 63°C after the fermentation stage resulted in a >4.5 log cfu/g reduction in the contaminating pathogen, although it could be detected by enrichment up to day 8 of the drying process. In pepperoni heated to 53°C for 60 min after the fermentation stage a similar >4.5 log cfu/g reduction was noted, although the presence of the organism could be detected to the end of the process (day 18).

Salamis and raw dried meats are highly traditional products with the processes of manufacture being difficult to modify for associated reasons. Key to the safety of these products is an understanding by the manufacturer of the factors inherent in the process that affect microbial growth and survival. Studies to determine the changes occurring throughout the process in controlling factors such as pH reduction, acidity development, moisture loss, water activity decrease and the effect these changes have on the microbial population including *E. coli* O157 derived from challenge testing product (in experimental facilities) are strongly recommended. Understanding these will allow the normal process changes that are shown to control the hazards to be incorporated as routine process controls such that any process that does not subsequently achieve the required change in the allotted time can be highlighted for particular attention as the chances of pathogen survival may be greater in these batches. Many manufacturers incorporate tests for *E. coli* or Enterobacteriaceae in finished products as indictors of process efficacy and hygiene. Such testing can be a useful means of assessing potential process survival or post-process contamination but as some strains of *E. coli* O157 are known to be resistant to acidic conditions they are likely to survive the process better than the organisms that may be used as indicators, making them ineffective indicators for this hazard. However, a fermented meat product shown to contain high levels of Enterobacteriaceae or *E. coli* is clearly of significant concern as it does indicate ineffective processing or major post-process contamination.

Final product issues and control

Salamis are usually sold in bulk form, often after vacuum packaging, or they may be sliced and sold as prepacks. The final products have a very low water activity, usually <0.90, high aqueous salt levels (5–10%) and low pH (<5.0), although all of these factors vary considerably depending on the individual product type. Opportunities for contamination exist during the slicing and packing operations but control of personal hygiene practices and implementation of regular effective environmental and equipment cleaning and disinfection will reduce the opportunities for contamination with *E. coli*. An important consideration is the segregation of raw materials from finished product, including the personnel working in each area. The operation of an appropriate flow in the factory from raw material storage and processing through to fermentation, drying and final slicing and packing will minimize opportunities for cross-contamination arising. Preventing operatives who handle raw meat from also handling finished product is an obvious but critical control of such hazards to the finished product.

A further opportunity for contamination arises with bulk product supplied to retail delicatessen counters. Under such circumstances the control of cross-contamination from raw meats to finished ready-to-eat products like salami is critical. This may include storage and display in different cabinets and the use of separate utensils and slicers for handling the product. Control of personnel hygiene is again critical and training in basic standards of hygiene is essential to ensure that all food handlers are aware of the hazards associated with cross-contamination and can take appropriate care when serving customers.

Table 4.8 Survey of raw fermented and dry-cured meat products on sale in the UK (adapted from Anon., 1997b)

	Product						
	German salami	Italian salami	Pepperoni	Country ham	Cervelat	Dried meat	Unknown sample type
Number of samples	183	98	58	50	46	3	17
Enterobacteriaceae							
< 10 per gram	179	95	57	49	44	3	16
10 to < 10^2 per gram	2	2	1	1	1	0	0
10^2 per gram or greater	2	1	0	0	1	0	1
E. coli							
< 10 per gram	182	98	58	50	46	3	17
10–10^2 per gram	1	0	0	0	0	0	0
> 10^2 per gram	0	0	0	0	0	0	0
E. coli O157:H7							
Detected in 25 g	0	0	0	0	0	0	0

The shelf life of salamis varies from weeks to months and as no growth of *E. coli* O157 will occur during this stage this is not a significant consideration for control of the hazard. However, it is possible that as the organism appears to die out during extensive dry storage, longer shelf lives may actually make the products safer.

Surveys have been carried out to determine the incidence of *E. coli* O157 in fermented meat products but routine microbiological testing has rarely resulted in the detection of these pathogens. A major study of fermented meat products on sale in the UK conducted by the Ministry of Agriculture, Fisheries and Food (MAFF) (Anon., 1997b) revealed no detection of *E. coli* O157 in 455 samples (Table 4.8). Some samples did, however, contain Enterobacteriaceae and *E. coli*, indicating the potential for survival or contamination with enteric pathogens. The benefits of applying end product testing for a pathogen such as *E. coli* O157, where it is likely to be present at extremely low incidence and levels, has been the subject of some debate (Anon., 1996c). However, use of microbio-logical testing to monitor the quality of the raw material meat and selective testing of the final product may be important considerations in respect of processes involved in traditional salami production that clearly do not achieve the complete destruction of even moderate levels of contamination in the raw material.

RAW-MILK MOULD-RIPENED SOFT CHEESE

Raw-milk mould-ripened soft cheeses are, in theory, the highest risk animal-derived products in relation to *E. coli* O157 and other VTEC. However, it is surprising to note that, so far, no VTEC food poisoning outbreaks have been attributed to the consumption of this group of products, although VTEC have caused outbreaks associated with raw-milk fromage frais (Anon., 1994a) and two outbreaks, 1 involving 22 cases, were attributed to raw milk hard cheese (Anon., 1997c,; Sharpe *et al.*, 1995). Raw-milk cheeses form a traditional product group, their production methods having been developed over many years, and they are often manufactured in small farmhouse operations. In recent years, production methods have become more sophisticated and automated, with modern equipment and management methods. However, the traditional nature of their manufacture has been retained therefore the risk of foodborne pathogens being present, although perhaps less than in previous years, still remains.

Raw-milk mould-ripened soft cheeses are traditionally synonymous with French cheese and the most famous varieties include raw-milk Brie and Camembert, which are characterized by their white mould coat. The most famous blue-veined raw-milk cheese, Roquefort, again has its origin in France; to this day it is ripened in the caves of a single rockface in France.

Description of process

Raw-milk mould-ripened soft cheese is manufactured from raw milk (Figure 4.3) that may originate from a variety of animals, although the most common types are made with cows' milk. Roquefort is manufactured from ewes' milk and other cheese varieties may be made from raw goats' or even buffalo milk. It is important to recognize that the nature of the raw milk undoubtedly influences the frequency at which enteric pathogens like *E. coli* O157 may be present as the organism may be more prevalent in some animals, such as cows, than in others, such as sheep or goats, although this will probably vary just as much between different animals on the same farm as it may vary between the same animal species on different farms. Depending on the size of the creamery, raw milk is either supplied in bulk milk tankers or, if a farmhouse operation, it may actually be taken directly from the milking parlour for use on the same farm. Milk is usually processed within 48 h of milking, during which time it is kept under refrigeration at temperatures not exceeding 5°C, although legislation may encompass raw-milk temperature control in some countries. The manufacturing process for raw-milk cheese is no different to pasteurized milk cheese apart from the omission of a pasteurization step. The process involves prewarming of the milk, usually through a plate heat exchanger to approximately 30°C, followed by inoculation of the milk with starter culture bacteria. Milk is fermented for a short period prior to the addition of rennet to coagulate the protein. During manufacture the cheese may be continuously or intermittently stirred or left for the curd to form as a block. After several hours the curd is usually formed and then cut to facilitate whey drainage and the whey is drained off. Some cheeses are manufactured in moulds where the entire fermentation and curd formation occurs in the container instead of in large vats. Following coagulation of the milk protein the fresh cheese is formed by filling the curd into moulds of particular shapes. The size and thickness of the cheese mould is one of the key factors that differentiates many of the cheese types like Brie, Camembert and Coulommiere. After filling the moulds, they are usually stacked on top of each other and left at ambient temperatures for 1–2 days to allow further drainage to occur and for the fermentation to continue. During the early stages of fermentation the pH of the milk decreases from approximately 6.5 to below 5.5 because of the growth of starter culture bacteria; this will also result in an increase in organic acid production, predominantly in the form of lactic acid. After forming the cheese in the mould the whole cheese round is removed

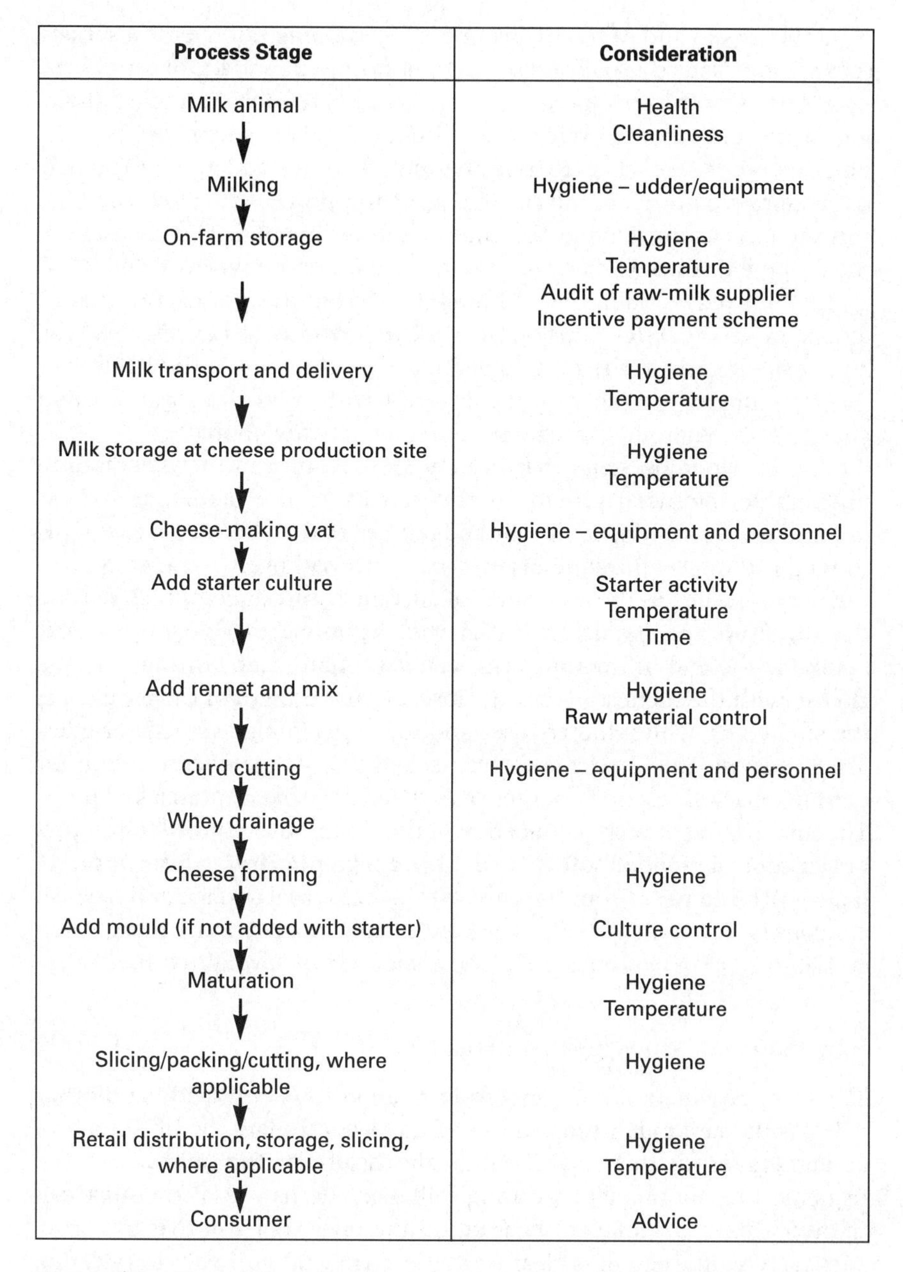

Process Stage	Consideration
Milk animal	Health Cleanliness
Milking	Hygiene – udder/equipment
On-farm storage	Hygiene Temperature Audit of raw-milk supplier Incentive payment scheme
Milk transport and delivery	Hygiene Temperature
Milk storage at cheese production site	Hygiene Temperature
Cheese-making vat	Hygiene – equipment and personnel
Add starter culture	Starter activity Temperature Time
Add rennet and mix	Hygiene Raw material control
Curd cutting	Hygiene – equipment and personnel
Whey drainage	
Cheese forming	Hygiene
Add mould (if not added with starter)	Culture control
Maturation	Hygiene Temperature
Slicing/packing/cutting, where applicable	Hygiene
Retail distribution, storage, slicing, where applicable	Hygiene Temperature
Consumer	Advice

Figure 4.3 Process flow diagram and technical considerations for a typical raw-milk mould-ripened soft cheese.

from the mould and salted by brining or by rubbing salt over the surface (dry salting). This is usually followed by spraying the surface of the cheese with spores of the specific mould species required for the mould ripening of the outside of surface-ripened cheeses like Camembert or Brie. Blue-veined cheeses usually have the mould spores added with the bacterial starter cultures at the beginning of the process to allow the blue veins to appear throughout the cheese during ripening. The cheeses are then ripened for several weeks at temperatures between 8 and 12°C under controlled conditions of humidity. During this period the characteristic mould coat develops on the surface of white-rind cheeses and the blue veins appear in the blue-veined cheeses, although in the latter case this is facilitated by piercing the cheeses with spikes to form aeration channels throughout the cheese along which the mould grows. It is during the ripening stage that the cheese becomes most vulnerable to undesirable microbial growth as the activity of the mould, as well as imparting a characteristic flavour and texture to the cheese, raises the pH to within the growth range of most bacterial pathogens. The activity of the mould results in the formation of alkaline by-products, which elevate the pH from approximately 5.0 at the beginning of ripening to near neutral at the end of ripening. This can vary significantly throughout the cheese with the highest pH being closest to the growth of the mould, i.e. the surface on white-rind cheeses and the veins in blue-veined varieties. After ripening, the cheese is either packed as a whole cheese round for cutting on a delicatessen counter or is sliced and sold as prepacked units. The final products vary significantly in their microbial controlling factors but, in general, ripened soft cheeses have a high pH varying from approximately pH 5 to pH 7 from the centre to the external surface and have an aqueous salt content of 1-5%. They are usually retailed under refrigerated conditions (<5°C) and may be given a shelf life of several weeks.

Raw material issues and control

The main raw material of concern in relation to *E. coli* O157 and other VTEC is the raw milk itself. The key factors controlling the likelihood of contamination at this stage include the health of the animal and the hygienic precautions taken during milking. The nature of the milk can therefore have a significant bearing on the degree of risk that the products carry with them. It is clear from the variety of outbreaks implicating bovine products that the greatest hazard is likely to be presented by cheeses made from raw cows' milk. However, VTEC have been isolated from a wide range of milking animals, including cows, sheep and goats (Beutin *et al.*, 1993), and therefore raw milk from any source must be considered to present a hazard with respect to the occasional presence of

VTEC and *E. coli* O157. The frequency at which VTEC will be present in milk is not widely reported but it is likely that it will be an infrequent contaminant. Padhye and Doyle (1991) reported a 10% (11/115) incidence of *E. coli* O157:H7 in raw milk, although a large study of raw milk, dairy and associated samples in the UK failed to detect the organism (Neaves *et al.*, 1994). A recently published survey of unpasteurized milk on sale in the UK found *E. coli* O157 in 3/1097 samples (de Louvois and Rampling, 1998).

Contamination from the animal is most likely to arise in two areas: first, from infection in the udder and shedding of organisms into the milk and, second, from faecal contamination of the external surfaces of the udder. Although it is possible for *E. coli* to cause mastitis (Bramley and McKinnon, 1990) this is unlikely to be a frequent occurrence in relation to human pathogenic types such as VTEC. Such incidences are likely to arise from faecal contamination of the udder, leading to pathogens entering the teat and causing an opportunistic infection. Mastitis of this nature can obviously be limited by appropriate animal husbandry practices that monitor the health of the animals and provide conditions under which such sources of infection can be reduced. In fact, many of the practices employed to prevent ingress of faecal contamination from the external surfaces of the udder into the milk during milking are also likely to have beneficial effects in reducing the potential for faecal contaminants to become opportunist pathogens invading the teat and udder and causing mastitis. The primary area for attention in reducing the potential for contamination of milk with VTEC is therefore the reduction of faecal contamination of the udder. Areas directly contacting the udder, including bedding and milking parlour equipment, are of particular importance. Foremost among the controlling factors is effective removal of faecal material from the udder prior to milking, using appropriate decontamination procedures of the udder and employing sanitizer solutions. It is often the obvious factors that are most frequently overlooked but, if they are controlled effectively, they can be the most effective in reducing the hazard. However, it is not possible to ensure sterility of the udder prior to milking therefore with increasing incidence of VTEC in herds and it is likely that on occasion these enteric pathogens will enter the milking parlour and raw milk. Under such circumstances a key factor in minimizing contamination of raw milk will be the effective control of the hygiene of the milking equipment. Effective cleaning and sanitization regimes conducted at regular intervals are critical to ensuring any contaminating pathogens do not colonize and build up in the milking equipment providing foci for extensive and high level contamination of further milk batches. In addition, effective temperature control of the milk will prevent growth of most contaminating bacteria but clearly a key to success for control of pathogens capable of causing infections at low doses is to minimize the frequency of occurrence as well as controlling growth. Areas requiring particular attention include the teat cups, which provide areas that may trap contaminants, together with the in-line strainers/filters and the milk-receiving vessel. Areas where product residues may build up, e.g. 'dead-legs' of pipes and tanks, need to be identified and removed to facilitate effective cleaning and sanitization regimes. In the absence of good equipment design and construction and effective hygiene standards it can be expected that contaminants entering the equipment may serve to contaminate future batches of milk and may even contaminate the cows themselves during milking.

Although small farmhouse enterprises are often viewed as being technically inferior to larger automated manufacturers, it is often the farmhouse manufacturer who is best positioned to control pathogenic microorganisms entering the raw-milk supply. This is because many farmhouse cheese manufacturers control both the supply of raw milk, coming from their own herds, and the production process itself. As the most critical point of this process is the raw-milk quality, direct control over the hygienic standards in operation is a distinct advantage. In situations where direct control of the raw-milk supply is not possible it is critical to the safety of the product that strong relationships are built up with the milk-supplying co-operatives and indeed the supplying farms themselves. Suppliers of raw milk to manufacturers of raw-milk cheese should be made aware of the end use of their milk and should be aware of the critical role that the hygienic operation of their farms plays in the safety of the finished products.

Many manufacturers of raw-milk cheese operate incentive payment schemes for their farmers based on the results of monitoring the hygienic status of the incoming raw milk. Indicators of contamination such as Enterobacteriaceae or *E. coli* may be monitored at frequent intervals in samples from each farm and from bulk milk tanks, and payment increased where good control is demonstrated and reduced for poor control. Some cheese manufacturers exclude supplies from farms for extended periods if milk quality falls below agreed criteria. In this way farms supplying raw milk to a consistent high standard receive greater rewards than those not. This focus of attention on rewarding quality helps to maintain consistent high standards while encouraging farms to invest in hygiene and contributing the money to do so. A similar incentive payment scheme was operated in the UK by the Milk Marketing Board; this was based on levels of bacterial contamination and indicators of infection such as the somatic cell count. This resulted in a significant improvement in the bacteriological quality of milk in the UK (Bramley and McKinnon, 1990).

Studies to date of the incidence of *E. coli* O157 have rarely found evidence of extensive contamination in raw milk. This is not surprising as tests for pathogens that are likely to be intermittent contaminants, present at low levels and at a low frequency, will rarely yield a positive result. However, the presence of the pathogen must be anticipated given the nature of the raw material and the ease with which non-pathogenic *E. coli* gain access to the raw-milk supply. Indeed, perhaps the greatest concern in relation to these products is that any increase in the incidence of the pathogen in dairy herds will expose current processes and practices to greater frequencies and levels of *E. coli* O157, with an increased possibility of future devastating outbreaks.

In addition to raw material contamination it is also important to recognize that raw milk is often transported to manufacturing sites in bulk milk tankers and then stored in milk silos. Clearly, significant opportunities exist for build-up of contamination and cross-contamination to subsequent batches of raw milk. Hygienic design and construction of the transportation and storage equipment together with effective cleaning and sanitization regimes are critical to breaking any chain of contamination that may build up from one batch of contaminated raw milk. Particular attention should be paid to 'dead spots' that may exist in the tankers or in milk silos, such as manway lids and seals, milk inlet and outlet ports and flexible hoses used to connect the milk tank to the silo. Upturned pipes, redundant sample ports or valves poorly welded on the inside of the tank create focal points where cleaning efficacy may be compromised by poor plant design. It is also important for operatives not to become com-placent regarding the hygiene requirements for handling raw milk. An untrained individual may see little danger in poor operating practices for a material that they may consider to be already contaminated, thereby making the potential hazards even greater. Equipment used in the transport and manufacture of these types of product needs to be subjected to frequent inspection and monitoring for any evidence of poor cleaning efficacy such as 'dead spots' or shadowing, which may occur due to blocked or poorly positioned spray balls of automatic cleaning systems. Monitoring cleaning efficacy using indicators of hygiene, such as coliform bacteria or Enterobacteriaceae, or rapid indicators of hygiene, such as ATP bioluminescence techniques, can be useful additions to standard process checks. Such monitoring may be by swabbing selected areas or taking samples of final rinse water from tanks after cleaning. It is important to note that the design of the equipment and operation of effective cleaning schedules are more important than reliance on the results of hygiene tests and it is easy to become led by microbiological test results when, in fact, they should be used only as part of an integrated hygiene programme. Successful hygiene monitoring is dependent on the ability to gain access to areas where contamination may be building up. This is often not those areas that can be reached for taking swab tests. For this reason, the use of visual plant inspection can be much less expensive and, in some circumstances, may be equally effective as microbiological testing.

Clearly raw milk is not the only raw material used in the manufacture of raw-milk cheeses; two further important raw materials warranting control are rennet and the starter culture. Rennin (commercially available in rennet) is a digestive enzyme and is extracted from the abomasum of calves. Rennet also contains pepsin and salt. These days much rennin is derived from fermentative culture systems but potential hazards associated with the use of animal sources need to be recognized and appropriate controls implemented by the raw material supplier. These, in turn, must be subjected to monitoring as part of the supplier quality assurance programme. The other critical raw materials are the starter and mould cultures and as well as ensuring that they have the requisite activity necessary to perform the essential fermentation and ripening reactions it is also important to ensure that the cultures are not subject to contamination from other hazardous microorganisms. Again, supplier quality assurance programmes that include assurance of the systems of hygienic production supported by microbiological examination for indicators of hygienic manufacture should be part of the intake requirements operated by the cheese manufacturer.

Process issues and control

Following receipt of the raw milk, the safety objective of the process used to manufacture raw-milk mould-ripened soft cheese must be to limit the potential for growth of any contaminating pathogens and prevent cross-contamination occurring. The key to controlling the growth of pathogens present in the raw-milk is the activity of the starter culture bacteria introduced at the beginning of the fermentation stage. Small manufacturers generally utilize starter cultures purchased from major starter culture manufacturers in which the activity and hygiene is quality controlled to high standards. Such cultures are usually supplied in dried or frozen form and are used to prepare the inoculum for the milk fermentation. Larger manufacturers may operate large-scale starter culture preparation tanks, which are usually contained in separate rooms to ensure the highest standards of hygiene are maintained. Clearly the manufacture of bulk starter cultures needs to be carefully controlled to prevent contamination from any environmental pathogens and also to prevent build-up of bacteriophage capable of infecting the starter cultures themselves, reducing their activity and hence the rate of subsequent fermentation during cheese manufacture. Monitoring for evidence of starter culture inhibition by assessment of culture activity is usually a component part of quality systems to ensure the starter is adequate for use.

Inoculation of the starter culture into prewarmed milk marks the beginning of the fermentation stage and it is the growth and acid production by these bacteria that have a significant influence on the control of contaminating pathogens. As temperatures of approximately 30°C are maintained for extended periods during the fermentation, it is clear that inadequate activity of the starter cultures could allow extensive growth of any contaminating pathogens that may be present in the raw milk. Rapid development of acidity during the fermentation is critical to the ultimate safety of the product and this is usually accepted as a critical control point for monitoring in any cheese manufacturing process, especially if the cheese is made from raw milk. During the fermentation of a typical cheese the pH usually drops moderately during the first few hours from approximately 6.5 to 5.5 or below, and may continue to drop during the following days of production to below pH 5.0. The presence of increased acidity and the lower pH is likely to have an inhibitory effect on the growth of *E. coli* O157 and other VTEC. However, it is likely that this effect will, at best, prevent the growth of the organism rather than result in any significant reduction in numbers. The acid tolerance of some VTEC strains is likely to have major implications for products that rely almost entirely on the antimicrobial effect of pH and acidity for the control of contaminating enteric pathogens.

Studies conducted on the survival of *E. coli* O157 in the manufacture of cheddar cheese have shown that levels may increase by over 1 log cfu/g during the initial stage of fermentation (Reitsma and Henning, 1996) and it is likely that similar growth would occur in the manufacture of raw-milk mould-ripened soft cheese. It is important to note that levels of *E. coli* O157:H7 in hard cheddar cheese actually decrease significantly by several orders of magnitude during the subsequent maturation stages of the process and it is probably the combination of decreased moisture, high salt content, acidity and pH that act to decrease the contaminating pathogens. Such factors are not present in the raw-milk mould-ripened soft cheese process and therefore the best that can be expected is for the fermentation stages to preclude the significant growth of any contaminating pathogens, should they be present in the raw milk.

Although unlikely to represent a major source of VTEC, it is important not to overlook the fact that contamination during the cheese-making process may occur from other sources in the environment. Contamination from equipment that was contaminated with pathogenic *E. coli* from inadequately treated river water was postulated as a cause of an outbreak of enteropathogenic *E. coli* in soft ripened cheese (Marier *et al.*, 1973) and therefore the control of contaminants introduced from the environment is important. Again, high standards of equipment design and construction together with effective cleaning and sanitization regimes will reduce the opportunities for cross-contamination to occur from the environment. Some cheese-ripening facilities are notorious as areas where contamination from the environment may occur and be extensive. Maturation rooms are usually maintained under temperature and humidity conditions that are within the growth range of bacterial pathogens and, as mould-ripening of the cheese develops, changing product characteristics may allow such growth of contaminating pathogens to be supported due to elevation of pH and reduction in acidity as a result of mould metabolism. Some cheeses, e.g. Roquefort, are ripened in caves on wooden shelves where the control over environmental contaminants is difficult to achieve. While this is of greatest concern in relation to pathogens such as *Listeria monocytogenes*, the potential for extensive contamination of the environment and subsequent transfer to products during ripening needs to be considered and appropriate environmental control implemented. Particular attention should be given to product contact surfaces, including shelving, and emphasis must be placed on efficient cleaning to remove any product debris and effective sanitization regimes to decontaminate surfaces. A further hazard in relation to any raw-milk cheese process is associated with the potential for contamination of finished product with raw milk due to inadequate segregation of raw and finished product production areas. Such control should be an integral part of the design of the process flow. Operator practices need to exclude those operatives handling raw milk from handling finished products and appropriate factory and process design considerations and staff training must be implemented to control this potential hazard.

Final product issues and control

Raw-milk mould-ripened soft cheeses are susceptible to the growth of bacterial pathogens in the finished product following the elevation of pH by the mould during ripening. In fact, it is likely that the key controlling factor for enteric pathogens in the finished product is storage temperature, as enteric pathogens such as *E. coli* O157 are reported not to be capable of growth at refrigeration temperatures (<5°C) although growth has been reported in foods with high levels of general microflora at temperatures above 8°C (Palumbo *et al.*, 1997). Under conditions of poor temperature control (10°C), growth may be possible in the finished product given the long shelf lives of these products. The reason for such concern regarding the safety of these types of cheese is evident. Few challenge tests have been conducted with VTEC in raw-milk mould-ripened soft cheese. However, work has been documented on the growth and survival of other pathogenic *E. coli* strains in Camembert cheese. Frank *et al.* (1977) demonstrated a clear difference between the potential for pathogenic *E. coli* to survive and grow depending on the point at which contamination occurred in or on the cheese. When inocu-lated into the milk used for the manufacture of Camembert cheese, *E. coli* initially increased by 100-fold in the first 6 h but then survived poorly during the process, reducing by 10–100-fold within 48 h to 1 week and then decreasing by over 10 000-fold over the ripening stage (3–7 weeks). Survival was actually better in the centre of the cheese than near the mycelial growth, where conditions would be expected to be less harsh. Importantly, when *E. coli* was inoculated onto the mycelial mat of the cheese during ripening, as would occur with post-process contamination, extensive growth ensued during the subsequent ripening stage. Levels actually increased by 1000-fold and the organism remained at a high level throughout the 7-week maturation period. Such findings indicate the potential consequences for post-process contamination of all types of mould-ripened soft cheeses, whether made from raw or pasteurized milk, and reinforce the need for high standards of personnel and environmental hygiene at all times. Park *et al.* (1973) studied the survival of enteropatho-genic *E. coli* during the manufacture of Camembert cheese, reporting growth during the first 5 h of fermentation from levels of approximately 10^2 per gram to 10^4 per gram. As the pH decreased during storage overnight to levels below pH 5.0, the level of *E. coli* also began to decrease by 10- to 100-fold and during maturation the organisms became undetectable between 1 and 9 weeks. The effect of antibiotics in the milk was also studied to determine whether inhibition of the starter culture enabled enhanced growth and/or survival of contaminating enteropatho-genic *E. coli*. The presence of penicillin (0.3 units per ml of milk) inhibited the rate of fermentation and prevented the cheese reaching a pH value below pH 5.0, allowing *E. coli* levels to increase by 6 log cfu/g within the first 24 h of fermentation. These levels then remained fairly constant throughout the subsequent maturation period, although levels decreased slightly (10-fold) between week 4 and week 8.

Studies on the survival of *E. coli* O157 have been conducted in cottage and cheddar cheese. Arocha *et al.* (1992) inoculated *E. coli* O157:H7 into milk used for the manufacture of washed curd cottage cheese. The process involved fermenting the milk at 32°C from an initial pH of 6.6 to approximately 4.7 within 5 h, with a concomitant increase in acidity from approximately 0.4 to 0.8%. The curd was cut and cooked to 57°C over a 1.5h period and then cooled by washing with water at room temperature followed by chilled water (4°C). Levels of *E. coli* O157:H7 increased by 100-fold during the fermentation stage of this process even in the competitive and inhibitory conditions created by the lactic starter cultures, i.e. low pH and high acidity development. Once the temperatures reached 57°C, death of the organism was reported and it could not be detected by enrichment procedures.

Reitsma and Henning (1996) studied the survival of *E. coli* O157:H7 during the manufacturing process of cheddar cheese. The organism was inoculated into pasteurized milk at a high (1000 cfu per ml) and a low (1 cfu per ml) level and this was used for the manufacture of cheddar cheese. The cheese was fermented with starter cultures at 32°C for 30 min followed by addition of chymosin to coagulate the milk proteins. After a further 20–25 min the curd was cut and scalded at 38°C for 30 min prior to draining the whey and cheddaring the curd to a pH of 5.2–5.3 for 2–3 h. Curds were milled and salted prior to placing into hoops and pressing overnight at 21–22°C. The cheeses were then packed and ripened at 6–7°C. The final pH of the cheese ranged from 4.95 to 5.2 and the aqueous salt content ranged from 2.75 to 3.76%. Results from the initial manufacturing stages (up to pressing) demonstrated that *E. coli* O157:H7 could survive and apparently grow during this stage of the process although some of this increase will have been due to concentration effects in the process. Levels increased from 1200 cfu per ml in the milk to 27 000 cfu per ml in the curd after pressing. In the studies with the low inoculum levels starting at 1 per ml, these increased to 60 per ml in the curd after salting but declined to 5 per ml in the curd after pressing. In both cases, the organism was more concentrated in the curd than in the whey although levels in the whey after cutting (84 per ml for the high inoculum cheese and 0 per ml for the low inoculum cheese) were seen to increase in the whey after scalding (500 per ml for the high inoculum cheese and 10 per ml for the low inoculum cheese). During ripening the levels in the high inoculum cheese decreased steadily by approximately 10-fold up to 14 days with a further 10-fold decrease after 28 days. Levels then remained fairly constant up to 74 days and were still present in four out of the five replicate cheeses (40 per gram, 20 per gram, 21 per gram, present in 25 gram) after 158 days (Table 4.9). In the low inoculum cheese, levels were difficult to assess because of the low numbers present. However, even after 60 days of ripening three out of the five replicates had levels of 1 cfu per gram and the other two were

Table 4.9 Survival of *E. coli* O157:H7 during the manufacture and ripening of cheddar cheese (adapted from Reitsma and Henning, 1996)

Process step	High inoculum cheese					Low inoculum cheese				
E. coli O157:H7 (cfu/ml or cfu/g)										
Milk	1200					1				
Curd post cooking	20 000					16				
Whey post cooking	500					10				
Curd after pressing	27 000					5				
E. coli O157:H7 (cfu/g) (five replicate cheeses)										
Ripening day 14	1900	3000	2800	2800	3500	< 1 (−ve in 25 g)	< 1 (−ve in 25 g)	< 1 (−ve in 25 g)	13	1
Ripening day 28	700	200	250	260	315	< 1 (+ve in 25 g)	< 1 (+ve in 25 g)	< 1 (+ve in 25 g)	< 1 (+ve in 25 g)	< 1 (−ve in 25 g)
Ripening day 60	31	12	117	26	332	< 1 (+ve in 25 g)	1	< 1 (+ve in 25 g)	1	1
Ripening day 158	40	< 1 (+ve in 25 g)	< 1 (−ve in 25 g)	20	21	< 1 (−ve in 25 g)	< 1 (−ve in 25 g)	< 1 (−ve in 25 g)	< 1 (−ve in 25 g)	

+ve = positive, −ve = negative

positive for *E. coli* O157 in 25 g. All samples tested were again positive in 25 g after 130 days but none of the organisms could be detected after 158 days of ripening. Bachmann and Spahr (1995) studied the survival of *E. coli* in Swiss hard (Emmentaler type) and semi-hard (Tilsiter type) cheeses made from raw milk. The manufacturing process for both these raw milk cheeses follows a standard fermentation profile at 32°C with lactic starter cultures and coagulation of the milk proteins. However, both cheeses are subject to elevated temperature cooking of the curd to expel whey. The hard variety is cooked at 53°C for 45 min while the semi-hard type is cooked at 42°C for 15 min. The whey is drained and the cheeses are pressed for 24 h and then brined in a 20% brine solution for a further 24 h. The cheeses are then ripened at 11–13°C for 90 days. The cheese process develops an initial low pH of 5.2 after one day and this increases slightly to 5.5–5.8 after ripening. The moisture contents in the final cheese are 35% (hard) and 39% (semi-hard), with salt contents of 0.5 and 1.2% (not aqueous), respectively. *E. coli* levels of nearly 10^6 per ml of milk were reduced to 10^2 per gram of curd after cooking in the hard cheese process, and following the first day of ripening no *E. coli* could be detected (<1 per gram). In the semi-hard cheese, the levels increased by 1 log cfu/g after cooking and pressing but declined to <100 cfu per gram after 30 days of ripening and were not detectable after 60 days ripening (<1 per gram).

The results of these studies give some insight into the ability of pathogenic *E. coli* and *E. coli* O157 to survive and grow in the conditions prevailing during the early stages of cheese manufacturing processes. This provides a salutary warning that raw-milk mould-ripened soft cheeses, which have no effective pathogen reduction element in the process, are high risk products with respect to these organisms and may be particularly vulnerable to acid tolerant VTEC.

Raw-milk mould-ripened soft cheeses are usually sold as whole cheeses, which may be cut into wedges on the delicatessen counters at food retailers or cut and prepacked at the manufacturing site. Under such conditions, the control of cross-contamination during the cutting process is important, primarily to prevent spread of any contamination throughout production batches or to subsequent batches. Attention must be given to the control of procedures for effective cleaning and sanitization of the cutting equipment, which is known to be difficult to clean because of the high fat content and adhering properties of the cheese. Wherever possible, cutting equipment should be completely stripped down to allow effective cleaning of otherwise inaccessible areas, e.g. blades, blade housing. This applies to the cutting operation at cheese manufacturers and in retail outlets. It is probable that the control of contamination of cutting equipment, even knives, in retail delicatessen operations is generally poorer than that in the manufacturing industry and it is essential that staff with responsibility for cleaning are aware of the hazards that are being controlled by their cleaning efforts.

Many factory and delicatessen cheese-cutting operations cut and pack a variety of products on the same equipment. It is important to be aware of the potential hazard of cross-contamination from unpasteurized cheese to pasteurized varieties in any cutting operation. In the factory, raw-milk mould-ripened soft cheeses are usually cut at the end of the production day so this can be followed by a full cleandown. On a delicatessen counter, however, because cutting often takes place throughout the day these cheeses should be cut on separate cutting equipment or using dedicated utensils. Care must be taken on retail displays to prevent raw-milk cheeses coming into contact with other products.

Surveys for the presence of pathogenic *E. coli* in foods have rarely reported detection of hazardous strains. However, it is apparent that *E. coli* generally is a frequent contaminant of raw-milk cheese. A survey published by the Public Health Laboratory Service of England and Wales (Nichols *et al.*, 1996) found high levels of coliforms and *E. coli* in a number of cheeses (Table 4.10). The type of *E. coli* found was not reported as tests for specific pathogenic strains were not carried out. In a study of 60 cheeses on retail sale in Iraq, Abbar (1988) reported the presence of *E. coli* in 43 samples with coliform counts ranging from 500 to 26 000 per gram. In addition, four of the *E. coli* strains were identified as pathogenic serotypes (O119:K69, O125:K70, O86:K61 and O111:K58). The nature of the cheeses (raw milk or pasteurized milk) is not clearly documented.

The fact that *E. coli* is such a common contaminant in raw-milk cheese occasionally present at very high levels gives clear warning that the processes used do not eliminate the hazard, if present, in raw milk. In fact, it is clear that with levels of *E. coli* in raw milk usually at <100 per ml and, more frequently, at <10 per ml, levels in excess of 100–1000 per ml in the finished cheese products must be representative of growth and concentration or extensive contamination during processing. If such contamination with apparently harmless types of *E. coli* were to be replaced by *E. coli* O157 or other VTEC then the consequences would be felt severely by those consuming the products.

Manufacturers of raw-milk mould-ripened cheese need to be clear that having the best controls in place to ensure high quality raw milk, control

Table 4.10 Levels of coliforms and *E. coli* found in a survey of cheese sold in the UK (adapted from Nichols *et al.*, 1996)

Levels	< 10 per gram	$10–10^2$ per gram	$10^2–< 10^3$ per gram	$10^3–< 10^4$ per gram	$10^4–< 10^5$ per gram	$10^5–< 10^6$ per gram	$10^6–< 10^7$ per gram	$> 10^7$ per gram
Raw-milk cheese (72 samples)								
Coliforms	40	5	3	9	2	4	9	0
E. coli	55	10	4	1	0	1	1	0
Pasteurized milk cheese (405 samples)								
Coliforms	297	38	19	23	9	9	9	1
E. coli	389	9	4	1	0	1	1	0

of fermentation and maturation processes and prevention of cross-contamination during processing is absolutely essential for minimizing the risk of foodborne outbreaks involving this most serious hazard. However, it must be recognised that this can only achieve a reduction in the risk of causing an outbreak. Even the best controls cannot preclude the strong possibility that cheese made from raw milk, especially the mould-ripened soft varieties, will eventually be the source of a major food poisoning outbreak involving VTEC. It is therefore essential to advise the public of the risks associated with the consumption of this type of cheese. In the UK it is now established practice for some major retailers to label their own brand products with advice that 'raw milk may contain organisms hazardous to health' and to point out that groups vulnerable to infection should avoid consumption of these types of cheese. It is apparent that governments are frequently reluctant to offer similar advice to the general public for fear of creating general public concern or receiving criticism from the industry fearing a backlash regarding safety. However, when faced with such evident and increasing hazards, such advice is surely to be expected in anticipation of future outbreaks and incidents. Precedents already have been set for this type of advice in relation to listeriosis from these types of cheese, however, this only arose after serious foodborne illness had already occurred in large outbreaks. In today's age of hazard analysis and risk assessment approaches it is clear that the public need advice in advance of such outbreaks to enable them to make informed choices about the foods that they consume. Proactive advice is surely more relevant than waiting for outbreaks of foodborne illnesses to occur before obvious hazards are given public airing.

Concerning the safety of hard cheese made from raw milk, it is clear that the lower moisture content and the retention of high acidity act together in contributing to the inhibition, decline and potential death of contaminating enteric pathogens. The pH of cheese such as raw-milk cheddar or even some of the continental cheeses like Gruyere or Emmenthal remains fairly low (pH 5.0) and with such low moisture content (35–40%), any inhibitory factors such as salt or other preservatives are concentrated in the aqueous phase where the microorganisms will predominate. Although work has demonstrated that *E. coli* O157:H7 levels are reduced during a cheddar cheese manufacturing process, it is interesting to note that two of the documented outbreaks of *E. coli* O157 food poisoning from raw-milk cheese were believed to have been caused by raw-milk hard cheese manufactured in the UK (Curnow, 1994; Anon., 1997). It is not clear however whether the contamination arose due to post-process recontamination or whether the organism actually survived the process.

FRESH-PRESSED FRUIT JUICES

Fresh-pressed fruit juices have been involved in several foodborne outbreaks of illness due to *E. coli* O157. These products represent a major part of the fruit juice market, constituting the premium products of the market. Products may include many varieties of fruit but the market is dominated by fresh orange and apple juice. The products have a strong 'health' association and the fact that many are fresh and unpasteurized adds to customer appeal. These products have a limited shelf life because of the presence of contaminating spoilage microorganisms, such as yeasts, which compromise quality and can cause spoilage within several days. They are sold under refrigeration and consumed by all sectors of society.

In the UK it is common practice to refer to fresh-pressed juice from apples as 'juice' whereas in the USA the term 'cider' is applied to distinguish it from pasteurized further processed apple juice. The term 'cider' in the UK is reserved for the description of fermented apple products that contain high quantities of alcohol. This has led to some confusion about the nature of the product and the extent to which the organism *E. coli* O157 represents a danger in fermented products. To date, all the outbreaks relating to apple juice ('cider' in the USA) involve fresh-pressed apple juice that has not undergone a heating or fermentation process.

Description of process

The production of fresh-pressed juices involves a fairly simple process. The raw material fruit is collected from the orchard and then transported to the producers of the fruit juice (Figure 4.4). The fruit is not necessarily of the highest grade as that is usually reserved for sale as fresh whole fruit. The fruit may be washed in water, which may sometimes be chlorinated, prior to entering a press. The actual method of juice extraction does vary depending on the fruit but, in most circumstances, the product is subjected to external pressure to break down the fruit cells, releasing the juice into a receiving vessel either directly or via extraction pipes. The juice may be filtered, although fresh juice often contains a high quantity of suspended solids, including some fruit pulp and is maintained under chilled conditions until filling.

To maximize the quality and shelf life of the product the juice is usually filled on the day of extraction into plastic or glass containers and distributed to retailers with a shelf life of usually less than 7 days.

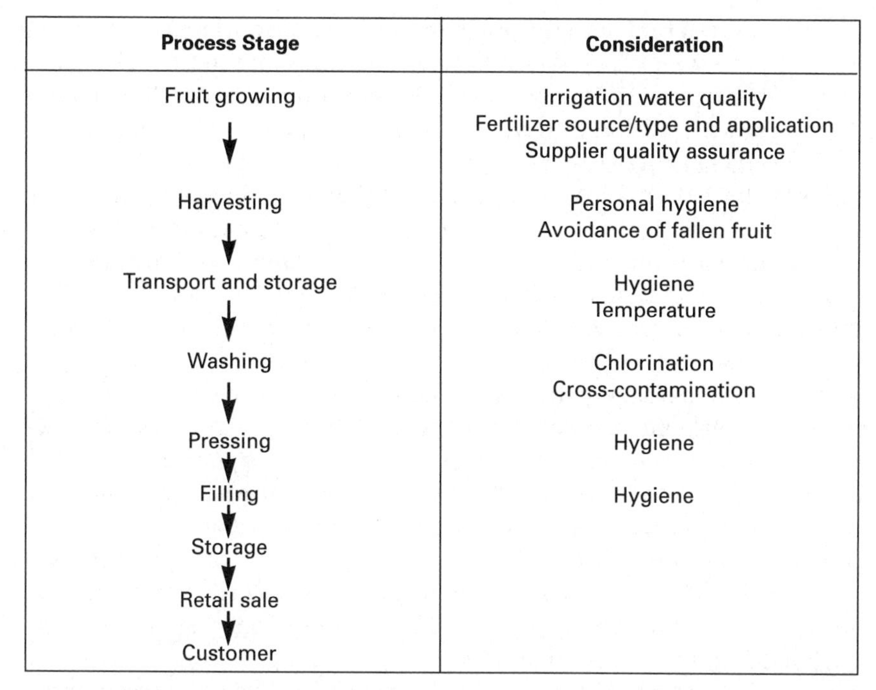

Figure 4.4 Process flow diagram and technical considerations for a typical fresh-pressed apple juice.

Raw material issues and control

As the fruit juice process has very few stages, starting with the fruit and ending with the bottled, extracted juice, it is clear that the primary area for control must be with the raw fruit itself. Apples, which are the only fruit implicated in outbreaks so far, are collected from orchards after hand picking. The fruit is stored at low temperature to minimize mould growth and then transported to the juice producer. The fruit is naturally subject to contamination during growth in the orchard. The soil may receive nutrients from artificial fertilizers or from animal waste in the form of compost or manure. In addition, fruit trees require water, which may be supplied from ground irrigation channels or from aerial irrigation. The potential for contamination of the fruit itself is therefore evident as microbial pathogens are likely to be present in the immediate surrounding environment. Contamination is only likely to occur when the set fruit is exposed during a fairly short ripening period of 2 to 3 months. Although it is possible for pathogens to contaminate the fruit in dust or via contamination from insects and defecating birds on the trees, such sources are unlikely to be major routes of contamination. Wallace *et al.* (1997), however, recently reported a high incidence of *E. coli* O157 in 800 samples of fresh bird faeces from gulls, lapwings, crows and jackdaws collected from a waste disposal landfill site and tidal sands in the UK. A total of 0.9% of bacterial isolates from samples from the urban landfill site and 2.9% intertidal sediments were contaminated with Vero cytotoxin-producing *E. coli* (1.6% overall incidence). While it is not known exactly how the birds became carriers, it is postulated that such birds feed on waste tips, farm slurry and sewage outlets, where they acquire the organism. The hazard such birds then present to the environments that they subsequently frequent, such as farmland, is evident but the extent to which this presents a hazard to fruit growing in orchards is not known. Most of the outbreaks in the USA have pinpointed one factor as the most likely source of contamination: the use of fallen apples. In circumstances where the orchard ground is subject to contamination by animal waste, the practice of picking fruit from the ground represents a significant contamination hazard to the fruit, but it is also a practice that may be extremely difficult to eliminate. It is human nature to want to make the best use of resources available and every apple on the ground is one more to use. However, if fallen apples are contaminated by animal wastes on the ground, the contamination potential for the production batch of juice is obvious. Following the apple juice food poisoning outbreak in 1991, a survey was conducted of fresh-pressed apple juice manufacturers in New England, USA to determine the extent of the practice of using fallen apples. All of the 36 respondent companies used 'drop' apples in fruit juice production (Besser *et al.*, 1993).

There is no doubt that animal waste will, on occasion, be contaminated with faecal pathogens, including *E. coli* O157. Surveys of the incidence of *E. coli* O157 in cattle and dairy herds have usually involved sampling the fresh faeces or examination of rectal swabs. Although the incidence appears to vary significantly due to seasonal and shedding variations, it is clear that an incidence of nearly 1% can be expected (Hancock *et al.*, 1997a; Gibbens and Wray, 1997). Once faeces are shed onto the ground *E. coli* O157 can survive for extended periods. Maule (1997a) studied the survival of *E. coli* O157:H7 in laboratory-scale ecosystems and demonstrated only a 2 log cfu/g reduction in 54 days in cattle faeces and a 1 log cfu/g reduction in soil cores after 63 days. A significant reduction was observed in cattle slurry, however, with a >5 log reduction in 9 days; similar findings were reported in river water (Table 4.11). Further work reported by Maule (1997b) indicated a significant difference in the survival of *E. coli* O157 in soil depending on the nature of the soil conditions. A 1–2 log reduction was observed after 130 days in soil cores with rooted grass, whereas a >5 log cfu/g reduction was recorded within 25 days in sieved soil.

Table 4.11 Survival of *E. coli* O157:H7 in simulated environmental ecosystems (adapted from Maule, 1997a)

Day	Cattle faeces (per g)	Cattle slurry (per ml)	River water (per ml)	Soil cores (per g)
0	7.1×10^7	1.2×10^8	6.7×10^6	8.1×10^7
1	9.7×10^7	1×10^9	-	-
7	-	1×10^4	3.8×10^3	2.2×10^8
9	2×10^7	ND	2.8×10^3	-
13	-	-	5.5×10^1	-
14	1.5×10^7	-	-	9.3×10^7
34	5.7×10^6	-	-	2.8×10^7
54	3.8×10^5	-	-	-
63	-	-	-	8.7×10^6

ND = not detected.

Under conditions of controlled application of treated animal waste it should be possible to prevent high levels of pathogen contamination . For example, composting of manure for several months can result in high temperatures being achieved in the compost and with regular turning an effective 'pasteurization' of the manure can be achieved with such a material being suitable for application to the ground. The safest alternative would be the use of artificial fertilizers where nutrient input can be carefully controlled and contamination contained. In some of the outbreaks it is reported that contamination may have arisen from faecal matter from cattle grazing in the orchard and while this certainly presents a hazard, the access of grazing cattle to fruit orchards is unusual as the trees would be subject to significant disturbance by the presence of large animals. If animal wastes are to be used, it is clear that no wastes should be applied when the developing fruit is exposed as contamination is far more likely during these stages. The best options for control are the application of artificial fertilizer and prohibition of the use of fallen apples. It is important also to control the quality of irrigation water supplied to the orchards as this can be a source of many enteric pathogens, particular in poorly developed countries. The key areas for attention and control are therefore:

- use of animal manure
- irrigation practices
- access to grazing cattle
- collection of fallen apples.

Fruit juice producers should be aware of these issues and an assessment of them should be built into any hazard analysis of their processes. As with raw-milk cheeses and fermented or dried meats, the fruit juice producer is largely dependent on third parties for the application of controls to help maintain the safety of the product. The development of a partnership approach with the raw fruit suppliers is likely to be critical for ensuring the consistent supply of high quality, clean fruit to prevent the presence of enteric pathogens in the raw material. It should be recognized that fresh-pressed fruit juice is often a premium product whose shelf life is mainly limited by spoilage due to growth of yeast contaminants. As most contamination arises from the raw fruit itself, it is evident that use of poorer quality fruit will compromise the quality and shelf life of the product as well as compromising its safety. Fallen fruit are more likely to be damaged and more susceptible to infection by moulds and yeasts. A policy to prevent the supply of poor quality fruit to minimize the presence of enteric pathogens is just as likely therefore to also control yeasts, which are responsible for high product losses. In addition, it is important to ensure all fruit collection and transport containers are kept clean and in good repair to minimize any cross-contamination that may occur between loads.

Process issues and control

With the application of effective controls in fruit-growing areas it is anticipated that contamination of fruit by enteric pathogens is likely to be infrequent. However, it is possible to further minimize the potential hazard by implementing effective controls during further processing. Fruit used for the production of fresh juice is usually selected (graded) and then washed prior to extraction of the juice. The selection is primarily designed to remove any obviously contaminated products and spoilt (mouldy), damaged or infected fruit from entering the process. The washing stage is in place to further reduce any foreign body contamination such as soil or stones. If contamination of fruit by enteric pathogens has occurred, it is likely to be present in discrete pockets with most fruit being free from contamination and some fruit being contaminated with perhaps high levels of pathogens. Fruit washing has the potential to distribute such contaminants across the whole batch if bulk wash tanks are used or if water flotation systems recycle the water. Cross-contamination during washing can be avoided by employing discrete washing systems such as spray washing but whichever system is employed it is important that consideration is given to the use of chlorinated water. The use of chlorinated washing water is important primarily to prevent the water itself being the source of any contamination or, in bulk systems, to prevent the spread of contamination throughout the batch. Levels of free chlorine in excess of 10 ppm are likely to be sufficient to prevent extensive contamination but if any decontamination of the fruit is required, levels in excess of 100 ppm may be needed.

Washing, even with chlorinated water, is recognized as not being a totally effective mechanism for eliminating bacterial contamination from fruit and produce but it can reduce contamination by one or two orders of magnitude (Adams *et al.*, 1989).

After washing, the fruit is subjected to juice extraction and, as batches of fruit are processed through common equipment, e.g. presses, juice extraction units, bulk storage tanks and filling lines, high standards of cleaning efficacy and sanitization are needed to prevent cross-contamination from batch to batch. This should include clear breaks in processing for daily cleaning or, depending on the circumstance, cleaning and sanitization between batches of fruit. The build-up of pulp and other fruit debris in and on equipment must be attended to and the processing environment must receive particular attention. Areas where such build-up occurs may need to be stripped down for cleaning.

Final product issues and control

Once VTEC has been introduced to the product from raw material fruit or during juice extraction and filling no further process activities will eliminate the hazard. The product is chilled and sold with a limited shelf life of 5 to 7 days, although this is dependent on the control of yeast contamination levels during the process. Attention to effective hygiene procedures can pay dividends in reduced spoilage by preventing build-up of yeasts as well as resulting in a safer product.

The survival of some VTEC strains in acid conditions has already been discussed and it is this acid tolerance that has emerged as a key factor in outbreaks of *E. coli* O157 foodborne illness attributed to these high acid products. Outbreaks to date have only affected apple products and have not been reported in fresh-pressed orange juice. While this may reflect differences in their processing and perhaps fewer opportunities for external contaminants to enter the juice, it may be that survival is greater in apple juice because the acidity is conferred by malic acid rather than the citric acid in orange. Although little data have been published in this area, the antibacterial effect of malic acid may be postulated to have a less adverse effect on *E. coli* than citric acid although the relative concentrations of acid and the pH of the fruit juice may also favour survival in apple juice over orange juice. Data have been reported by various authors on the survival of *E. coli* O157 in fresh apple juice. Miller and Kaspar (1994) investigated the acid tolerance and survival of *E. coli* O157:H7 in apple cider (juice in the UK) and in trypticase soya broth (TSB) adjusted to low and high pH. Two *E. coli* O157:H7 isolates were compared to a control strain

Table 4.12 Survival of *E. coli* O157:H7 (log cfu per ml) in trypticase soya broth under acidic conditions (adapted from Miller and Kaspar, 1994)

		pH		
Strain	Time (h)	2	3	4
E. coli O157:H7	0	4.8	4.9	5.0
	1	4.8	4.9	4.8
	7	4.6	4.9	5.0
	24	3.8	4.8	4.8
Control *E. coli*	0	3.5	4.4	4.5
	1	2.8	4.5	4.4
	7	< 1.0	4.3	4.3
	24	NT	2.8	4.0

NT = not tested.

of *E. coli*. In all cases the control strain was much less acid tolerant than the strains of *E. coli* O157:H7 when inoculated into apple cider and stored at 4°C or when inoculated into TSB at pHs ranging from 2 to 12 and stored at 4°C and 25°C (Table 4.12). The presence of the preservatives potassium sorbate (0.05 and 0.1%) or sodium benzoate (0.1%) had little impact on the survival of *E. coli* O157:H7 in apple juice after storage at 4°C for 21 days (Table 4.13). In a study over 14 days at 4°C the initial levels of one of the *E. coli* O157:H7 isolates remained unaffected and decreased by less than 0.5 log cfu/ml during the experiment in unpreserved apple cider, apple cider with 0.1% potassium sorbate and apple cider with 0.1% sodium benzoate. In contrast, the control *E. coli* strain decreased by approximately 2 log cfu/ml within the first 3 days in all products and was undetectable (4 log decrease) between day 5 and day 7. A second *E. coli* O157:H7 isolate also decreased but at a significantly slower rate than the control strain with nearly a 4 log cfu/ml decrease occurring after 11–14 days. Interestingly, for this second strain levels decreased faster in the unpreserved apple juice than in those containing preservatives. The research concluded that under normal conditions of storage *E. coli* O157:H7 present as a contaminant in

Table 4.13 Survival of *E. coli* O157:H7 in apple juice containing preservatives (adapted from Miller and Kaspar, 1994)

Type of preservative	pH	Survivors (%)*
Potassium sorbate (0.1%)	3.9	90
Sodium benzoate (0.1%)	4	43
No preservative	3.9	91

* Percentage survivors after 21 days at 4°C in apple juice containing preservatives.

the fruit juice will not be destroyed within the shelf life of fresh-pressed apple cider. Zhao *et al.* (1993) conducted similar work with apple cider (apple juice in the UK) containing preservatives and found contrasting results; their strain of *E. coli* O157:H7 actually decreased significantly faster in preserved juice. This may be attributable to the strain of *E. coli* O157:H7 used in the studies and/or the effect on *E. coli* O157 of the mould growth that occurred during some experiments. In apple juice stored at 8°C with pH values between 3.6 and 4.0 initial contamination levels of approximately 10^5 per ml in six production lots increased slightly and then remained stable for 12 days, with one exception in which levels decreased by 4 log cfu/ml within this time. Levels in all lots then gradually declined over the successive 36-day storage trial. Levels of *E. coli* O157:H7 became undetectable after 15, 20 (two lots), 28 (two lots) and 34 days. These differences were attributable to variations other than the pH of the products. When cider was inoculated at initial contamination levels of 10^2 per ml and stored at 8°C, levels remained between 10^2 cfu per ml and 10 cfu per ml for 8 days in all products and became undetectable (<5 cfu per ml) between 11 and 14 days. In contrast, when inoculated into the six lots of apple cider at levels of 10^5 per ml and stored at 25°C, levels increased by 1 log cfu/ml in two lots and decreased by 1 log cfu/ml in a further two lots within 2 days. Levels became undetectable between day 3 and day 6, demonstrating the enhanced effect of elevated temperatures on the destruction of the organism. This is likely to be related to the enhanced effect of organic acids on microbial death at elevated temperatures.

The presence of potassium sorbate (0.1%) had little effect on the survival of *E. coli* O157:H7 compared to the control apple cider. Sodium benzoate (0.1%), however, had a statistically significant effect on enhancing the destruction of *E. coli* O157:H7 at 8°C. In all but one lot the initial contamination levels were reduced to undetectable levels (>4 log cfu/ml reduction) within 7 days at 8°C and a similar enhanced reduction was seen at 25°C with levels reduced to undetectable numbers within 2 days in five lots and 3 days in the remaining lot. The combined effect of sodium benzoate (0.1%) and potassium sorbate (0.1%) was also reported to enhance antimicrobial activity at 8°C by reducing survival times by 50% when compared to sodium benzoate alone. This did not occur, however, at 25°C. Semanchek and Golden (1996) studied the survival of *E. coli* O157:H7 in fermenting and non-fermenting apple cider. In fermenting apple cider initial contamination levels of log 6.4 cfu per ml were reduced by less than 1 log cfu/ml within 24 h, by nearly 4 log cfu/ml after 2 days and to undetectable levels (*c.* 6 log cfu/ml decrease) after 3 days at 20°C. The pH of the product started at 3.62, reducing to 3.48 at day 3 and ending at 3.72 after the 10-day incubation period. Alcohol content (% vol/vol), however, changed significantly with levels increasing from 0 (day 0) to 0.63 (day 1), 1.94 (day 2), 3.03 (day 3) and ending at 6.01 at day 10. The enhanced effect of the process in eliminating *E. coli* O157:H7 was attributed to the production and antimicrobial effect of ethanol. In the non-fermenting apple cider levels of *E. coli* O157:H7 remained within 1 log cfu/ml of the initial level of log 6.5 cfu per ml for 7 days and then decreased steadily to log 2.9 cfu per ml after the 10-day incubation. The alcohol content of the product remained at undetectable levels and the pH changed little, starting at 3.62 and increasing slightly to 3.75 (similar to the fermented product) after 10 days.

To make apple juice safer in relation to *E. coli* O157 and other VTEC some processors have introduced product pasteurization. Research by Splittstoesser *et al.* (1996) has shown that *E. coli* O157:H7 is highly sensitive to mild temperatures in apple juice. The mean D_{52} value was 18 min with a z value of 4.8°C. Heat resistance was unaffected by increasing the Brix from 11.8 to 16.5° but was reduced by increasing the l-malic acid concentration from 0.2 to 0.8%, reducing pH from 4.4 to 3.6 and by the presence of the preservatives sorbic acid and benzoic acid (Table 4.14).

Few surveys have been published on the incidence of enteric pathogens on fruit or in fruit juice. Monitoring for *E. coli* O157 or other enteric pathogens is probably of little value in a product whose safety should be achieved and maintained by effective raw material and process controls. Some information may be gained by monitoring levels of indicators of enteric contamination such as *E. coli* in the fresh-pressed juice immediately after pressing, before the effect of acidity on the indicator organisms causes a reduction in numbers. However, as such indicators are likely to

Table 4.14 The effect of different preservation factors on the heat resistance of *E. coli* O157:H7 in apple juice (adapted from Splittstoesser *et al.*, 1996)

Preservative factor	Preservative level	D_{52} (minutes)
°Brix	11.8	23
	16.5	22
% l-malic acid	0.2	24
	0.8	15
pH	3.6	14
	4.4	35
Sorbic acid	200 mg/l	36
	1000 mg/l	5.2
Benzoic acid	200 mg/l	22
	1000 mg/l	0.64

be less acid resistant than *E. coli* O157 itself, the usefulness of end product surveys for such contamination may be limited.

The application of combined strategies to prevent spoilage and reduce the incidence and survival of pathogens in fresh fruit juices will be important in making these products safer. The use of preservatives however appears to have a questionable effect on product safety in relation to *E. coli* O157 and would also diminish the appeal of these products to the consumer.

COOKED MEAT PRODUCTS

Cooked meat products form a major food commodity group and are consumed by all sections of the population. Cooked meats do not have a good history in relation to outbreaks of foodborne illnesses and examples can be found of outbreaks caused by most types of food poisoning bacteria in association with this group of products, including *E. coli* O157 (Stevenson and Hanson, 1996; Gammie *et al.*, 1996). Cooked meats may be derived from a variety of raw materials but the most popular are pork, poultry and beef. As a variety of cooked meat products may be manufactured in the same processing plant the species of meat being used becomes less important as bacterial contamination may well be spread extensively during the raw material processing itself.

Description of process

Cooked meats are manufactured from any type of food raw material meat: pork, beef, lamb, chicken, turkey and many others. The process should be one of the simplest to control as it consists of few stages that, if controlled effectively, should result in little, if any, enteric pathogen contamination of the final product. Unfortunately, the degree of sophistication in large-scale manufacture has made factory operations much more complex and offers significant opportunities for post-process contamination, whereas in small-scale manufacture the problems with control of cooking efficacy and raw/cooked separation remain key factors in the continued hazard presented by this group of commodities.

Cooked meat products are manufactured by cooking bulk meat as a joint or in a comminuted and reformed state in an oven, achieving temperatures sufficient to destroy most vegetative microbial contaminants and effecting changes in the textural properties of the meat to provide organoleptically acceptable product (Figure 4.5). The temperatures usually applied in the UK for the cooking of meats are consistent with the guidelines given by

Process Stage	Consideration
Animal husbandry	Health Cleanliness
↓	
Animal slaughter and processing	Hygiene Temperature
↓	
Meat transport, delivery and storage	Hygiene Temperature
↓	
Comminuted and Reformed bulk meats Bowl chopping and addition of other ingredients (spices, herbs, salt, etc.)	Hygiene Temperature
or	
Whole joints of meat Brine injection (where applicable, e.g. hams and cured meats)	Hygiene Temperature
↓	
Cooking	Temperature Time High/low risk segregation (post cooking)
↓	
Blast chilling	Hygiene
↓	
Removal from container (where applicable)	Hygiene
↓	
Storage	Hygiene Temperature
↓	
Roasting/chilling, where applicable	Temperature
↓	
Super chilling	Hygiene
↓	
Slicing, where applicable	Hygiene
↓	
Garnishing	Raw material control Hygiene
↓	
Packing	Hygiene Temperature
↓	
Storage/distribution	Temperature
↓	
Retail storage	Hygiene Temperature
↓	
Retail slicing, where applicable	Hygiene
↓	
Retail sale	Hygiene Temperature
↓	
Consumer	

Figure 4.5 Process flow diagram and technical considerations for a typical cooked sliced meat.

the Department of Health (Anon., 1992a). This indicates that a safe process can be achieved by subjecting all parts of the meat product to a process of 70°C for 2 min. Following cooking, the bulk meat is usually blast chilled to reduce its temperature quickly in order to prevent the growth of any surviving spore-forming bacteria and then it is stored chilled at 0–5°C. Bulk meat, cooked in hermetically sealed containers, may be sold to retailers for slicing on delicatessen counters and under such conditions the meat, if cooked, cooled and handled correctly, is subject to little contamination prior to being received, sliced and sold at the retail outlet. The majority of cooked meats, however, are subject to some form of further processing, which usually involves removal from the container in which it was cooked and then chilling prior to roasting and/or slicing. Any cooked meat that is further processed receives significant handling and exposure to environmental contamination therefore control of these factors is critical to the safety of the products. Cooked meats are usually sold with a shelf life of between 1 and 4 weeks at chill temperatures depending on product type, although this is of little relevance with respect to *E. coli* O157 as the organism would not be expected to grow under reasonable refrigeration conditions (<8°C) during manufacture, distribution, retail and consumer storage (Palumbo *et al.*, 1997).

Raw material issues and control

The raw materials used for cooked meat products vary significantly depending on the final product required. The raw meat itself may be treated with salt, emulsifying salts (often by brine injection), herbs and spices. Few other antimicrobial compounds are added to the meat although sodium nitrite may be included in formulations containing pork and it is the inclusion of nitrite that differentiates ham products from cooked pork. The nitrite is added to give a characteristic flavour and colour to ham but it also confers some antibacterial activity predominantly against spore-forming bacteria such as *Clostridium* species. Nitrite is rarely added to meats other than pork.

The greatest hazard associated with cooked meat products comes from the raw material meat used. The incidence of *E. coli* O157 and other VTEC in meat animals has been reviewed in previous sections and it is clear that they will, on occasion, be present in raw meat entering the processing plant. The extent of contamination is dependent on the meat species with beef being the predominant species of concern, although its presence in other meats such as lamb and pork has been reported and should be expected (see Table 1.6). Minimizing pathogen contamination of the raw material coming into the processing plant can only be achieved by those practices employed at the supplying slaughterhouse and every effort must be made to assure high standards of hygienic processing operated by the raw meat supplier. This may be supplemented by microbiological monitoring of the raw meat as part of a supplier quality assurance programme to provide information regarding the hygiene of the supply. It is also important to recognize that contamination can be spread extensively in the processing plant after receipt if effective handling, cleaning and sanitization regimes are not implemented. Although the cooked meat process involves an effective heat-processing stage, spreading the hazard from discretely contaminated carcasses to other raw meats in the processing plant increases the opportunities for further cross-contamination thus elevating the risk associated with the final products. Operating practices should be employed that eliminate opportunities for extensive cross-contamination and prevent areas where contaminants may build up on the raw meat side of the processing unit. Processing equipment such as mincers, bowl choppers, brine injectors, conveyors and all other product contact surfaces need to be cleaned and sanitized daily with interim cleandowns during production to prevent the build-up of contaminants on equipment. Such practices need to be extended to all areas of the processing environment.

Process issues and control

The most critical stage in any process involving cooking is the heat process (time, temperature) used. Effective cooking and control of post-cooking contamination will render cooked meat products safe in relation to vegetative enteric pathogens, including *E. coli* O157. Most processes employed in the cooking of meat exceed temperatures of 70°C. However, overcooking can result in an organlopetically inferior product and so processes are often established according to the minimum time/temperature combination required to achieve the appropriate flavour, visual and textural qualities while also ensuring the product meets the minimum requirement for the destruction of vegetative pathogens. In the UK the established minimum heat process applied to cooked meat products is 70°C for 2 min or an equivalent process. This heat process is capable of achieving a significant reduction in any contaminating VTEC. Work with naturally contaminated ground beef has shown that this process would achieve in excess of a 6 log cfu/g reduction in meat contaminated with *E. coli* O157 (Advisory Committee on the Microbiological Safety of Food, 1995). While it is possible to achieve a 6 log cfu/g reduction with a lower heat process, it is important to build in some degree of safety margin to account for any factors that may reduce the efficacy of heat processing, e.g. variations in initial product temperature.

Betts *et al.* (1993) studied the heat resistance of *E. coli* O157:H7 in pork and chicken homogenate, demonstrating a *D* value at 64°C of 0.46 min with a calculated *z* value of 7.3°C. Orta-Ramirez *et al.* (1997) reported the *D* value in ground beef at 68°C to be 0.12 min (Table 4.15). It is therefore important for cooked meat manufacturers to fully understand any limitations of the cooking processes employed for the specific products handled.

Orta-Ramirez *et al.* (1997) also examined the thermal resistance of *E. coli* O157:H7 in ground beef in an attempt to identify an intrinsic enzyme within beef tissue that was destroyed according to similar inactivation kinetics. It was concluded that the thermal inactivation properties of triose phosphate isomerase was similar to *E. coli* O157:H7 and therefore this enzyme has potential as an indicator of effective processing for destroying contaminating strains of the organism.

Having established the heat process limits required to ensure the safe destruction of contaminating vegetative pathogens, the heat process must be carried out such that all parts of the meat product receive the full process. This must take account of the different positions in which the product may be placed in the oven together with the different sizes and types of meat that may be cooked.

Validation studies are an essential prerequisite for the establishment of safe cooking processes for raw meat products. Meat is often cooked in ovens where control is normally of the oven temperature and time of cook. Control of these process parameters is acceptable, providing they are based on trial studies demonstrating achievement of appropriate times and temperatures for each product type. Oven process controls do not take account of the product thickness, density, temperature of ingoing raw material and variability in heat distribution in the oven. These must form part of the temperature validation study to determine the cold spots in the oven, and the process times and temperatures established should take account of the worst case situations, e.g. using the thickest product. Once established, each cook must achieve the validated cooking profile and the process should be monitored with a continuous temperature recorder. Following the cook, the centre or 'cold spot' temperature should be taken of products known to be in the coolest parts of the oven. The temperature should be taken by inserting the sterilized metal probe of a temperature measurement device into the deepest part of the meat.

It is essential that any equipment used for monitoring such a critical part of the process is calibrated to ensure accuracy of measurement. It was surprising to learn therefore that in a recent survey of cooked meat

Table 4.15 Heat resistance of *E. coli* O157 in different meat substrates

Temperature (°C)	*D* value (min)	*z* value (°C)	Fat content	Inoculated product	Reference
53	46.1	5.59	3.8%	Ground beef	Orta-Ramirez *et al.*, 1997
58	6.44				
63	0.43				
68	0.12				
54	13.05–20.10	5.24 (7.29)*	Not reported	Chicken homogenate	Betts *et al.*, 1993
56	5.96–7.14 (5.19–5.49)*				
58	3.22–4.35 (2.20–2.38)*				
60	1.25–1.30 (1.15–1.62)*				
62	0.37–0.46 (0.67–0.75)*				
64	(0.36–0.46)				
54	15.56–18.48	4.98		Pork homogenate	
56	8.86–9.89				
58	3.75–3.95				
60	1.46–1.48				
62	0.42–0.45				

* Figures in parentheses are *D* values and *z* value obtained with a different strain of *E. coli* O157.

manufacturers conducted in the UK by the Ministry of Agriculture, Fisheries and Food (Anon., 1995e) 36% of processors did not calibrate their temperature monitoring equipment. Perhaps of even greater concern was the finding that a third of processors were unable to state the maximum temperature reached at the centre of their cooked meat products during the cooking stage. These factors in themselves do not make the products unsafe but it is ignorance of this nature that can be the forerunner of food poisoning outbreaks since not knowing the temperatures achieved during cooking indicates a lack of understanding of the need for the attainment of microorganism destruction temperatures. Also, there is the possibility that were safe temperatures not achieved, the processor would either not notice or not realize the significance of low cook temperatures. These days there can be no excuse for manufacturers of high risk products like cooked meats to remain ignorant and untrained in the technical requirements for controlling the safety of the products that they supply to the general public.

Providing the meat is given an effective cook, the only subsequent hazard to the product arises from cross-contamination during further processing. Meat from the ovens is initially cooled to dissipate heat and then blast chilled to reduce the temperature quickly. In some manufacturing units the products are cooled using shower systems where recirculated water is showered over the product to reduce the high temperatures, thereby facilitating subsequent rapid blast chilling. Meats may be cooked in hermetically sealed packaging or containers but if the seals of such containers or packaging are not sound then contamination can be introduced during subsequent cooling or chilling. It is highly unlikely that enteric pathogens will be introduced from the environment in a cooked meat operation providing the segregation between the raw meat and cooked meat areas is controlled effectively and personal hygiene practices are also effective. In most large-scale operations ovens used for cooking are 'through' ovens built into a dividing wall between the raw meat side of the factory and the cooked product side. The product is placed, by designated operatives, into the oven on the raw side of the unit. Sophisticated double-entry ovens ensure that while the doors on the raw side are open it is not possible to open the doors on the cooked side (and vice versa), thereby preventing the possibility of contamination passing through. Once the raw meat is loaded, the door is closed on the raw side and the process of cooking begins. Once cooking is complete, operatives dedicated to the cooked side of the factory open the door and the cooked meat is removed and subsequently handled in the cooked product area.

Effective barrier hygiene begins with the use of through ovens but must include complete control of the processing environment to prevent contamination being introduced by operatives themselves. Processing environments for the handling of cooked meats are often referred to as high risk or high care areas and the procedures in place to prevent contamination of product from personnel or the environment are extensive. Excellent guidance on the controls necessary in such areas have recently been revised and published by the Chilled Foods Association (Anon., 1997d).

The greatest potential hazards in relation to enteric bacterial pathogens arise from the raw meat itself and the hygienic practices of manufacturing personnel. Effective environmental hygiene practices are important to prevent the spread of any contaminants from the floors, walls, drains and other environmental areas to product contact surfaces and the product itself, but it is the raw meat and personnel practices that require particular emphasis on control measures. Controls of raw material meat have been discussed but control of personnel practices is a vital element in maintaining the safety of these products. Control of enteric pathogen hazards in relation to staff begins with a clear infectious disease policy where persons known to be unwell with enteric illness are not allowed into food factory environments. As anyone ill with *E. coli* O157 infection is unlikely to be at work anyway, because of the severity of the illness, the key stage for these persons is prior to their return to work. Bacteriological test clearance of stools for several consecutive samples needs to be considered before allowing anyone with a known illness of this nature to return to food handling environments (Anon., 1995a). In most circumstances, however, it is possible to control enteric pathogens by the implementation of effective personal hygiene procedures. This requires full training of operatives in food hygiene prior to allowing them to work in food handling environments. All personnel entering a high risk factory area should undertake a full change of external clothing, including replacement of shoes, coat and hair covering. Changing areas should be designed with a central bench and procedures that clearly separate the low risk side from the high risk area. Hands must be washed thoroughly prior to entry into the factory and similar precautions need to be taken before and after operatives visit the toilet. These simple precautions are usually sufficient to prevent personnel from transferring major food poisoning organisms to foods because of inadequate personal hygiene and, if implemented and followed properly, will be fully effective. It is essential that every individual entering high risk areas should follow the same precautions. Unfortunately it is often the case that personnel not properly trained in hygiene practices may break the rules and compromise safety. Typical examples include engineers brought in to fix plant breakdowns or senior management who believe that their level of seniority carries with it a hygiene barrier!

Cooked meat products may be stored in the packaging in which they were cooked or may be removed and the exposed product may be stored in the chiller. In these situations the product is exposed to both environmental and personnel contamination. Contamination of product by enteric pathogens is unlikely to occur in the chiller but attention to cleaning and the operation of high standards of hygiene in these areas are nevertheless important in controlling any build-up of general contaminants.

Some cooked meats receive a further heat process, e.g. roasting at high temperature (>200°C) to achieve a desirable visual appearance. While the surface temperature achieves a repasteurization, the centre of the product merely gets warmed to 30–40°C. The products are then rechilled. Operation of good hygienic practices is important throughout such additional processes during which product is exposed to potential environmental contamination.

The greatest opportunity for recontamination of cooked meat after processing arises during slicing and packing. Slicing product exposes large surface areas of the product to contamination from product contact surfaces such as meat loading equipment, slicing blades and conveyors used to transfer the sliced meat to a packing machine. Product contamination may also occur from people because of direct handling of the sliced product and by aerosols generated from non-product contact surfaces such as floors, walls or drains during cleaning. Although not likely to be a major source of enteric pathogens, the hygienic design and operation of slicing equipment is essential to minimize build-up of product debris and microbial contamination, and prevent extensive cross-contamination to subsequent products. Areas that require specific control measures include slicing equipment, conveyor belts, aerosols from the environment, cleaning processes, aerosols from forced air chillers, tables, knives, product storage containers and racks, and personnel handling practices.

Some cooked meat products may be garnished after cooking with herbs and spices for visual impact when on display on delicatessen counters. The potential for the introduction of enteric pathogens on garnishes needs to be carefully considered and appropriate control procedures implemented to prevent such occurrences compromising product safety. Where possible, garnishes should be treated by dipping in hypochlorite prior to introduction into the high risk area; alternatively, use of heat-treated garnishes should be considered whenever they can be sourced. Garnishes should be subject to some form of microbiological monitoring for enteric pathogens or indicators of contamination, e.g. *E. coli*.

It is important to point out that most of the controls described above use a large manufacturing environment as a model but many cooked meat products are manufactured under conditions where some of the controls discussed are unlikely to be possible. Many small operations, including butchers, may cook meat and areas used for the storage of cooked meat may be in the same room as those for the preparation of raw meat. In addition, such operations tend to be operated by few staff and therefore the possibility of dedicating staff for raw and cooked meat handling is not practical. Under these conditions, the controls necessary for ensuring prevention of cross-contamination are often inadequate and the potential for cooked meat contamination remains very high indeed. While the risk of these outlets causing extensive outbreaks is generally less than the larger national or multinational manufacturers or retailers, the outbreak in Scotland (Reid, 1997) clearly demonstrated how such small outlets can be associated with large and fatal outbreaks.

The control of product contamination by enteric pathogens in small production units requires the same principles as above to be operated. Procedures must be in place to prevent raw meat from coming into contact with cooked foods by direct contact, via people, via use of common equipment or by aerosol transfer. Cooking must be capable of achieving the destructive temperatures advised by the Department of Health and any further processing of cooked meat such as slicing should be conducted on dedicated equipment.

Final product issues and control

The presence of *E. coli* O157 or any other VTEC in a ready-to-eat cooked meat product is completely unacceptable and represents a significant health hazard to the general population. All control measures in place therefore must aim to eliminate the hazard being present as a result of either surviving the cooking process or by being introduced after the heat process from people or the environment. Once introduced onto a cooked meat product *E. coli* O157 levels are unlikely to change during the shelf life of the product if it is kept under appropriate conditions of temperature, i.e. <8°C. If the product is temperature abused the organism can be expected to actually grow given the formulation of typical cooked meat products. Cooked meat products sampled from the UK market had average salt content and residual nitrite levels of 2% and 12 ppm, respectively, with a pH of 6.1 and an aqueous salt content of <2.5% (Table 4.16), factors unlikely to restrict the growth of the pathogen, if present, under conditions of temperature abuse. The importance of temperature control is therefore clear to see although, of course, the organism should not be present in the

Table 4.16 Survey of cooked meats from retail stores: average levels of antimicrobial factors found (adapted from Anon., 1996d)

Type of meat	pH	Nitrite (residual) (ppm)	Aqueous salt (%)
Cooked cured pork	5.85	11.41	2.47
Cooked cured comminuted meat	6.12	12.89	2.42
Corned beef	5.94	-	2.2
Cooked poultry	6.11	-	1.7

first place. The presence of nitrite in some cooked meat products has very little effect on the growth or survival of *E. coli* O157. Under normal good conditions of temperature storage the shelf life of cooked meat products has little bearing on the levels of *E. coli* O157 or other VTEC which would survive and present a health hazard even at low levels.

Surveys of cooked meat products for the incidence of VTEC have failed to yield many positive isolations although contamination with *E. coli* and Enterobacteriaceae, as indicators of post-process contamination, is frequently detected. In recent surveys of cooked meat products conducted by the Ministry of Agriculture, Fisheries and Food in the UK, samples of cooked meat and poultry were examined for *E. coli* O157 and Enterobacteriaceae (Table 4.17). Samples were taken directly from manufacturing plants before slicing, after slicing and at the final hold stages of the process. In a further survey, an investigation assessed the incidence of contaminants in retail prepacked meat products (Table 4.18). *E. coli* O157 was not detected in 25 g samples taken from a combined total of 627 samples. However, Enterobacteriaceae were detected in a variety of products with levels increasing post slicing and on retail sale. Clearly, this provides evidence of the potential for cross-contamination during post processing although the high levels found in retail products may indicate psychrotrophic growth, possibly aided by mild temperature abuse.

The potential for cooked meat products to become contaminated during retail sale has received much attention following the outbreak in Scotland and the publication of the Pennington report (Pennington, 1997). Practices in retail stores aimed at preventing cross-contamination of raw meat to cooked foods vary significantly. However, it is clear that many raw meats are displayed in the same display counter as cooked, ready-to-eat foods and, in some cases, common utensils such as weighing scales are used. Indeed in the chillers where products are stored prior to sale,

Table 4.17 Survey of cooked meats for *E. coli* O157 and Enterobacteriaceae at secondary processing sites (adapted from Anon., 1996e)

Type of meat	Enterobacteriaceae (log cfu per g)								*E. coli* O157 (per 25 g)
	Total samples	< 1	1–1.99	2–2.99	3–3.99	4–4.99	5–5.99	> 6	
Cooked cured pork									
Presliced	17	14	2	1	0	0	0	0	0
Sliced	17	13	2	2	0	0	0	0	0
Final hold	17	13	1	3	0	0	0	0	0
Cooked cured comminuted meat									
Presliced	13	12	1	0	0	0	0	0	0
Sliced	13	10	3	0	0	0	0	0	0
Final hold	13	11	2	0	0	0	0	0	0
Cooked poultry									
Presliced	18	14	2	1	0	0	1	0	0
Sliced	18	13	3	0	0	1	1	0	0
Final hold	18	13	2	1	2	0	0	0	0
Corned beef									
Presliced	20	20	0	0	0	0	0	0	0
Sliced	20	19	1	0	0	0	0	0	0
Final hold	20	18	1	1	0	0	0	0	0

Table 4.18 Survey of cooked meats for *E. coli* O157 and Enterobacteriaceae from retail stores (adapted from Anon., 1996d)

Type of meat	Number of samples	Enterobacteriaceae (log cfu per g)									*E. coli* O157 (per 25 g)
		< 1	1–1.99	2–2.99	3–3.99	4–4.99	5–5.99	6–6.99	7–7.99	8–8.99	
Cooked cured pork	229	135	22	21	16	13	7	4	1	1	0
Cooked cured comminuted meat	47	34	3	5	2	1	1	0	1	0	0
Cooked poultry	99	42	4	10	7	8	11	13	4	0	0
Corned beef	48	16	5	5	9	5	4	4	0	0	0

storage conditions and handling practices may be even poorer as these areas are not open to public scrutiny. Reputable retailers have invested much time and money in introducing effective segregation practices both behind the scenes and in display cabinets. This is supplemented with extensive training of all staff in food hygiene requirements and investment in completely separate utensils, storage containers, slicing equipment and display counters for raw meats and ready-to-eat foods. Without such advancements in the implementation of effective segregation and hygienic operation further outbreaks of *E. coli* are inevitable as the incidence of this hazard increases in the raw material supply and the consequences of errors in food operations are severe.

Underpinning all of these safety measures should be a full and structured hazard analysis of the specific manufacturing operation. Application of informed common sense to the known hazards by focusing on the effective thermal destruction of pathogens and the avoidance of cross-contamination are the clear and overriding messages for this category of product.

RAW COMMINUTED MEAT PRODUCTS: BEEFBURGERS

Raw beef products have probably been responsible for the greatest number of reported outbreaks of foodborne illness attributable to *E. coli* O157 (Riley *et al.*, 1983; Wells *et al.*, 1983; Ostroff *et al.*, 1990; Belongia *et al.*, 1991; Bell *et al.*, 1994; Willshaw *et al.*, 1994; Advisory Committee on the Microbiological Safety of Food, 1995). In many countries beef is considered to be one of the most coveted of meats with premium products such as steaks often being a luxury item on the shopping list. In recent years, however, beef has been at the centre of a number of food scares and this has resulted in a decrease in its consumption in some countries, particularly in the UK. Of greatest concern have been the scares relating to a postulated link between the consumption of bovine material from cattle suffering from bovine spongiform encephalopathy (BSE) and Creutzfeld-Jacob disease (CJD) in humans. Together with nutritional concerns (saturated fat content) and outbreaks of food poisoning caused by *E. coli* O157 attributed to beef, some people have considered the risks of consumption to be greater than the pleasure they derive from eating the product. The products most commonly associated with outbreaks of *E. coli* O157 are beefburgers, which are considered to be a convenience food consumed by all sectors of the population but particularly cherished by young children.

Description of process

Cattle used for the production of beef are generally reared on farms for up to 3 years. The cattle generally graze on open pasture but also spend significant amounts of time in sheds, particularly in the winter, where they may be fed silage or artificial feed. Cattle may carry VTEC internally due to colonization of the gut and they may also carry the organism externally due to contamination of the hide from faecal discharges. Cattle are transported to slaughterhouses where they are received and held in enclosures prior to progressing to slaughter. After slaughter the animal is 'dressed', which includes removal of the hide and the viscera (stomach, intestines, etc.). The carcass is usually trimmed to remove any faecal contamination and may be washed with hoses to remove excess blood. It is then chilled over a period of several hours to achieve a temperature of <5°C. After chilling, the carcass may be further processed, either on site or after transport to further processors, by cutting it into primal joints, which are usually vacuum packed or packaged in gas flushed containers. These may then be stored for extensive periods at sub-zero temperatures (−2°C), frozen for shipment to further processing facilities or further processed on site. Further processing results in the production of individual cuts of meat, fillets, diced meats and mince for retail sale or processed products such as beefburgers. Beefburgers are manufactured from raw beef with the possible addition of fat, herbs, spices, crumb, salt and other flavourings (Figure 4.6). The raw materials are bowl chopped or minced to form fine pieces of meat and then reformed into meat patties of the required thickness and diameter. The products are packed and then chilled or frozen for despatch to retail outlets. In many cases beefburgers may be manufactured in retail outlets by mincing meat on site and forming the burgers in a similar way at the retail outlet but under such conditions the size of the burger are less well controlled. An excellent overview of the microbiology of raw meat as affected by processing techniques is given by the ICMSF (International Commission on Microbiological Specifications for Foods, 1998).

Raw material issues and control

The raw material of primary concern in relation to the hazard from *E. coli* O157 is, naturally, the raw beef meat. Contamination of the meat begins with the live animal itself and the practices in operation to minimize colonization and carriage in the gut. Clearly, *E. coli*, including O157 serotypes, can colonize the gut of the animal although it is evident that, with few exceptions, the organisms cause little or no apparent illness in

Process Stage	Consideration
Animal husbandry	Health Cleanliness
↓	
Animal slaughter and processing	Hygiene Temperature
↓	
Meat transport, delivery and storage	Hygiene Temperature
↓	
Bowl chopping and addition of other ingredients (spices, herbs, salt, etc.)	Hygiene Cleaning efficacy Temperature
↓	
Forming and packing	Hygiene Temperature
↓	
Storage/distribution	Temperature
↓	
Retail storage	Hygiene Temperature
↓	
Retail sale	
↓	
Caterer Consumer	Advice Cooking instructions

Figure 4.6 Process flow diagram and technical considerations for a raw beefburger.

the animal itself. Numerous studies have reported the incidence of *E. coli* O157 in animal populations, including dairy herds (Mechie *et al.*, 1997), cattle herds (Chapman *et al.*, 1993) and sheep (Chapman *et al.*, 1996). In a survey of bovine rectal swabs of animals at a local abattoir implicated as a source in outbreaks of human infection with *E. coli* O157 in South Yorkshire, UK, Chapman *et al.* (1993) reported 4% (84/2103) incidence of *E. coli* O157 of which 93% (78/84) were VT+. It is interesting to note that of animals where the bovine rectal swab was positive for *E. coli* O157, some 30% of their carcasses at the slaughterhouse were also contaminated with the organism whereas of the rectal swab negative cattle, 8% of carcasses were found to be contaminated with *E. coli* O157. In a similar survey of the incidence of *E. coli* O157 in rectal swabs of sheep immediately after slaughter, an incidence of 2.6% (18/700) has been reported (Chapman *et al.*, 1996). One of the most comprehensive studies conducted to date on the incidence of *E. coli* O157 in cattle was undertaken in the USA under the auspices of the USDA National Animal Health Monitoring System's Cattle on Feed Evaluation Project (Hancock *et al.*,

1997b). A total of 100 cattle feedlots were selected for study, covering cattle reared in a variety of states in the USA. Swab samples of faecal material were taken from fresh faecal pats (<2 h old) and 210/11881 were found to contain *E. coli* O157, representing an incidence of 1.8%. Of the 100 feedlots examined *E. coli* O157 was isolated from faeces from 63 with the prevalence within feedlots varying between 0 and 10%. Cattle that were on feed for shorter periods (average 7±11.1 days) had an incidence of *E. coli* O157 three times higher than those on feed for longer periods (average 185±69.4 days). In reviewing the factors likely to affect the incidence of *E. coli* O157 in the animals in feedlots, Dargatz *et al.* (1997) reported that type of feed was a significant factor, with barley increasing the risks and soymeal decreasing the risks. This may be due to a number of factors, including the extent of contamination of the feed itself or the effect of the feed on digestion and the subsequent effect on gut microflora dynamics. Increased likelihood of *E. coli* O157 carriage associated with shorter times on feed was attributed by the researchers to a number of factors, including stress due to recent introduction to the feedlot diet and increased colonization of cattle not previously exposed to contamination, i.e. *E. coli* O157 may be more readily detectable in the faeces of these newly colonized animals. Factors that were associated with decreased incidence of *E. coli* O157 were the weight of the animal on arrival at the feedlot, with animals greater than 700 lb (~320 kg) being less likely to have positive samples. This may reflect the fact that heavier animals may be older and therefore more immunocompetent. It is of some interest that factors not associated with increased shedding were the use of feeding antibiotics, animal density within pens, previous health status of the cattle and the use of coccidiostats. It is also important to note that the study merely examined the faecal shedding of *E. coli* O157 and, as shedding can occur intermittently, the results can only give a limited indication of carriage and not a total view of percentage colonization of the animals by *E. coli* O157 as many animals may have carried the organism without shedding.

The circumstances that have allowed *E. coli* O157 and other VTEC to colonize bovines and other animal species are not clear but it is possible that the organism has been selectively advantaged in some way, such as in its tolerance to acidic conditions, which may have allowed it to survive the harsh conditions in the gut and then compete effectively with other strains, both in the cattle gut and in the farm environment. The organism is clearly quite capable of surviving for extended periods at low pH values, which may favour its passage through the stomach, e.g. Miller and Kaspar (1994) reported the ability of *E. coli* O157:H7 to survive for extended periods at 4 and 25°C in tryptone soya broth in which the pH was adjusted with hydrochloric acid; levels remained fairly constant, decreasing by less than 1 log cfu/ml at pH 2.0 and pH 3.0 over a 24 h period.

During normal cattle grazing in fields it is unlikely that extensive cross-contamination between cattle would occur but it is probable that in the winter months, when cattle spend a lot of time in sheds, cross-contamination among animals may occur. The use of common cattle sheds and slurry tanks, however, may provide further sources from which the spread of *E. coli* O157 could occur. It is also possible that cattle may acquire the organism from the consumption of contaminated water as the organism has been traced to such sources on farms where infection has been identified. Additionally, the use of bovine manure on pasture for grazing cattle may also contribute to continuing the cycle of contamination.

However, in a recent survey of the incidence of *E. coli* O157 in 36 herds in the USA where 1.41% (179/12 644) of faecal samples were found to be contaminated, little difference between herds could be determined with respect to manure handling practices and incidence of the organism (Hancock *et al.*, 1997c). The incidence of *E. coli* O157 was similar in herds where the heifers were housed in dry lots or on pasture with or without application of manure. In addition, in this study the use of manure on land for forage crops was not statistically associated with the incidence of *E. coli* O157.

Some studies have indicated that feeding practices prior to slaughter may affect shedding of *E. coli* O157 during transportation, lairage and slaughter. Withholding feed may reduce the potential spread of gut contents during subsequent slaughter and carcass dressing, and therefore reduce the potential incidence of the organism on the carcass. Conversely, starving the cattle may actually increase stress and therefore lead to increased shedding of *E. coli* O157 (Tkalcic *et al.*, 1997). Work continues to be conducted to help the industry to gain a better understanding of the factors contributing to the incidence of the hazard in the animals during rearing and primary processing. This should ultimately benefit processing practices, allowing the organism to be better controlled.

Whatever the reason, it is clear that the organism can be carried for extensive periods in cattle and may be intermittently shed in faeces. Carriage by cattle may be affected by the pH in the rumen and it has been postulated that acidity in the rumen may affect the survival, colonization and shedding of the organism. An understanding of the sources and spread of the organism on the farm is essential to identifying ways of reducing its incidence by the introduction of relevantly targeted, best farming practices. However, it is inevitable that cattle being transported to slaughter will, on occasion, carry or be contaminated with *E. coli* O157. Under such circumstances, efforts must be focused on preventing its spread to other animals during transportation and prior to slaughter. Cattle are frequently heavily contaminated with faecal discharges on their hides and in many cases this can be in the form of a dry 'cake' that is laden with faecal contaminants. Such contamination can arise on the farm, in sheds, during transportation or in lairage. Stresses placed on the animal in the 'foreign' conditions of transport may increase faecal voiding and the potential for hide contamination. Minimization of dirty cattle arriving at slaughterhouses must be a critical point for industry control. This can be achieved primarily by good husbandry practices although, as indicated, it has also been suggested that starving the animal for a period before transportation may reduce the extent of faecal discharge. Such approaches must be balanced with the welfare requirements of the animal. Better designed transportation vehicles and even some form of washing or clipping of the animal to remove excess waste material from the hide prior to slaughter are further considerations in designing effective control strategies. In the UK, animals are inspected for evidence of disease by the Official Veterinary Surgeon (OVS) assisted by meat hygiene inspectors. Animals with extensive faecal contamination on the hide should be rejected by the receiving abattoir and guidance is provided on the degree of faecal contamination which is unacceptable. Enforcement of this aspect would clearly reduce the potential for dirty animals to contaminate other animals. Introduction of a incentive payment scheme for delivery (or supply) of clean animals to slaughter has been suggested as one approach to decreasing obvious visible contamination from the farm and during transport (Pennington, 1997). Certainly full and complete records should be maintained concerning the source and condition of animals to provide a key link in the traceability chain. Effective cleaning and high hygiene standards of the vehicles used for transportation are vital elements in the prevention of widespread cross-contamination between consignments of animals. Control of contamination is also required at the slaughterhouse where areas used to hold the animals prior to slaughter need to be managed effectively to avoid crowding and the spread of faecal waste. It is important to note that such areas also represent a significant opportunity for pathogen spread to other animal species as the lairage may be common to other animals although the slaughter lines themselves may be separate. This could account for cross-contamination of faecal pathogens like *E. coli* O157 to species of animals not normally associated with the organism, although it is far more likely that such cross-contamination will arise during further processing.

After slaughter it is essential to limit the opportunities for cross-contamination using hygienic slaughterhouse practices. The slaughter and dressing of cattle is never likely to be an aseptic process but it is essential to ensure that practices are in place that will prevent operatives from spreading contamination over the whole carcass and to other carcasses. Practices such as tying the anus and blocking the oesophagus to prevent faecal and other gut discharges have been investigated and may be useful methods for helping to prevent extensive cross-contamination. In addition, the careful removal of hides ('hide pulling') can reduce the contamination hazard from often heavily contaminated material. Some of the greatest cross-contamination hazards arise due to poor operator practices. Knives, protective handwear, 'chain mail', aprons and all product contact implements used by operatives, together with the washing practices for carcasses, can represent significant cross-contamination potential and regular cleaning and sanitization of all these items needs to be implemented to prevent this. Careful attention to the sanitization of knives, in particular, after cutting away areas of the hide prior to hide pulling will minimize opportunities for the transferral of enteric pathogens to the flesh. In addition, preventing cross-contamination via the hands of operatives in between handling the hide, the hoofs or other externally contaminated areas of the animal prior to handling exposed flesh will contribute to this goal. The use of high pressure cleaning of carcasses should not be employed; if water needs to be used this should be at low pressure to reduce aerosol formation. Any obvious faecal or soil contamination should be removed by trimming with knives. Again the emphasis in processing must be to minimize contamination of individual carcasses and, therefore, of subsequent carcasses. Following the review of the *E. coli* O157 outbreak in Scotland a full investigation was conducted by an independent committee who made some excellent recommendations regarding the control of the hazard in slaughterhouses (Pennington, 1997). Foremost among these was the implementation of a HACCP-based approach to the slaughter of animals, involving the adoption and implementation of common sense approaches to limiting the opportunity for contamination and cross-contamination. Some of the other points raised were:

- training of operatives in food hygiene
- provision of adequate space between carcasses on conveyor lines
- design of conveyors to avoid contact with walls and floors.

With the increasing concerns about the incidence of faecal contamination on carcass meat a number of alternative processing technologies has been investigated. An excellent overview of emerging decontamination techniques has been published by Corry *et al.* (1995). The technologies showing greatest promise are carcass-washing techniques and steam pasteurization. Carcass washing involves spraying or washing the carcass with an antimicrobial solution such as organic acids (acetic or lactic acid) or trisodium phosphate. Favourable results have been achieved with these solutions but other products such as those based on copper sulphate (pentahydrate) and N-alkyldimethylbenzylammonium chloride, respectively, were deemed to have no greater effect on reducing bacterial populations associated with faecal contamination on beef tissue than water washes (Cutter *et al.*, 1996). A promising technique for large-scale decontamination of bovine carcasses is steam pasteurization. The process, which has been completely automated and is contained within large compartments, involves a small number of carcasses entering a chamber in the steam pasteurization unit. The chamber is sealed automatically and the carcasses are exposed to pulses of steam entering the vessel at 90–94°C. The carcasses are held in the chamber for 6–8 s to create a localized 'pasteurization' at the surface of the carcass. The carcasses are then moved from the steam chamber to the cooling chamber where sprays of chilled water (<5°C) rapidly cool the carcass. In a trial conducted using a full-scale unit at a beef processing facility (Nutsch *et al.*, 1997), steam pasteurization resulted in a significant reduction in the microbial contamination level of the 140 carcasses studied. Prior to steam pasteurization, natural contamination of carcasses included *E. coli* at an incidence of 16.4% with levels ranging from 0.6 to 1.53 log cfu/cm^2 and Enterobacteriaceae at an incidence of 46.4% with levels between 0.6 and 2.25 log cfu/cm^2. After processing, the incidence of *E. coli* reduced to 0% and Enterobacteriaceae incidence reduced to 2.9% with levels between 0.6 and 1.99 log cfu/cm^2. In a comparison of steam pasteurization with other methods for reducing pathogens on cattle carcasses, steam pasteurization demonstrated a numerically greater pathogen reduction than trimming or hot water/vacuum spot cleaning alone (Phebus *et al.*, 1997). Trimming involves cutting out the contaminated area with a clean knife while vacuum cleaning involves use of a handheld device that forms a seal over a small area of the carcass and sprays hot water to decontaminate local areas of the carcass while creating a suction to remove the spray. In comparative studies, trimming, vacuum cleaning and steam pasteurization gave mean log reductions in *E. coli* O157:H7 of 3.1, 3.11 and 3.53 (Table 4.19). In contrast, use of water washing alone only decreased contamination levels by 0.75 log cfu/cm^2. It was demonstrated that even greater reductions could be achieved by implementing a number of measures together to decontaminate carcasses, such as the use of trimming, washing and steam pasteurization. Clearly, there is potential to significantly improve the hygiene status of carcasses by the application of some of these new technologies and it is likely that the meat industry will, in time, adopt some of these techniques.

Table 4.19 Effect of different carcass decontamination techniques in reducing levels of *E. coli* O157:H7 (adapted from Phebus *et al.*, 1997)

Treatment process	Mean reduction in *E. coli* O157:H7 (log cfu/cm^2)
Washing (35°C)	0.75 ± 0.49
Steam pasteurization	3.53 ± 0.49
Trimming	3.10 ± 0.49
Hot water/vacuum spot cleaning	3.11 ± 0.49
Trimming and washing (35°C)	4.71 ± 0.53
Trimming, washing (35°C), lactic acid spray (2%) and steam pasteurization	4.14 ± 0.53

Most processing is currently undertaken without the use of such specifically targeted decontamination systems and after primary processing of the carcass it is likely that contamination will be present in 'pockets' of contamination containing pathogens in certain areas of the carcass. It is during further processing that extensive spread of these hazards can also occur. Attention to regular cleaning and disinfection of knives and other product contact utensils and surfaces, such as conveyors and overhead conveying machinery, must be the focus for minimizing extensive spread of enteric pathogen hazards during processing. It is not possible for any text to detail all the areas that may exist within individual food processing environments that need to be effectively cleaned and decontaminated to prevent build-up of contaminants in the process environment. It is for this reason that a detailed hazard analysis is so important for each specific site because, when correctly applied, it can help processors to identify those areas that need particular control measures to maintain the safety of this most critical of materials.

The practice of co-processing different meats in secondary processing plants also needs to be considered as it is likely that contamination from beef could be spread to other meat species if cleaning procedures between the processes for different meat species are not employed. Assessing the hygienic status of meats by testing for bacterial indicators is a useful way of monitoring the trend associated with the hygienic manufacture and preparation of the meat.

Process issues and control

The greatest potential for spreading any contamination in raw material meats arises during the processing and manufacture of the beefburger itself. Although any hazards from other raw materials need to be considered, it is the raw meat that presents the greatest hazard to this process. The raw meat will, on occasion, inevitably be contaminated with faecal pathogens including *E. coli* O157 and therefore efforts must be focused to reduce the incidence and levels of contamination in the raw meat supply by effective supplier quality assurance systems. Once *E. coli* O157 has been introduced to the raw material, the process of manufacturing beefburgers will have little effect in reducing it. However, processing standards can influence the extent of cross-contamination between different batches. Every effort must be made to ensure that cleaning and sanitization of production equipment and the environment are conducted at regular intervals to prevent build-up of contaminants and thereby reduce the opportunity of spread to future batches of product. Bowl chopping equipment and mincing machinery need to receive particular attention, in addition to burger patty forming equipment. Emphasis must be placed on effective removal of product residues and then decontamination of the properly cleaned surfaces. Meat residues are difficult to clean effectively and expert technical help needs to be sought from cleaning chemical companies to support the specific needs of the manufacturing plant. Cleaning and sanitization efficacy should be supported by routine monitoring using traditional indicators of hygiene such as Enterobacteriaceae or rapid ATP bioluminescence hygiene monitoring.

The biggest problem in relation to beefburgers when considering the hazard of enteric pathogens such as *E. coli* O157 is the location of the pathogen in relation to the cooking method that will subsequently be applied. Pathogens introduced onto raw meat during primary processing will usually reside on the surface of meat joints and muscle cuts. Cooking applied by the consumer usually just needs to achieve decontamination at the surface of the meat joint in order to destroy surface contaminants therefore the effective destruction of pathogens on beef steaks and other cuts of meat is fairly simple to achieve. However, with comminuted products like minced beef the contaminants become evenly spread across a significantly larger surface area during mincing which, when reformed into a burger, results in pathogens being present throughout the product. There is very little that can be done in the process itself to prevent this from occurring and therefore the ultimate safety of the product will rely on the achievement of temperatures throughout the product sufficient to destroy contaminating pathogens during the final cook stage.

Final product issues and control

Beefburgers are either chilled or frozen for retail sale and such processes are unlikely to have a significant effect on the survival of enteric

Table 4.20 Survival of *E. coli* O157:H7 in frozen, ground beef (adapted from Doyle and Schoeni, 1984)

Storage at −20°C (months)	0	3	6	9
Number (cfu per g)	6700	3700	6600	6200

pathogens such as *E. coli* O157. Freezing may result in some reduction in the levels of the organism over a period of time but it cannot be relied on to achieve significant destruction of the pathogens. Doyle and Schoeni (1984) stored ground beefburgers inoculated with *E. coli* O157:H7 (Table 4.20) at −20°C for up to 9 months, during which time levels remained almost constant. Beefburger products will therefore be sold to consumers with occasional contamination with *E. coli* O157. It is therefore the caterer or consumer who operates the final critical control point in the process, i.e. effective cooking. Under such circumstances it is essential that the beefburger manufacturer provides information to consumers to allow them to cook the product safely. Such information is usually presented as cooking instructions on the packaging of the product. The formulation of a cooking instruction for a raw meat product like a beefburger is a complex task that must take account of the factors that influence the achievement of the cooking temperature that will make the product safe by destroying enteric pathogens. In all situations the instruction should be designed to achieve a cook sufficient to reduce initial contamination levels of *E. coli* O157 and other VTEC by at least six orders of magnitude. In the UK, the standard for beefburgers would be the achievement of 70°C for 2 min or an equivalent process (Anon., 1992a). In the USA advice varies from a 71.1°C internal end-point temperature for consumers cooking the beefburger to 68.3°C (holding for 16 s) for food service operations (Jackson *et al.*, 1996; Liu and Berry, 1996). Clearly, under either regime significant reductions in *E. coli* O157 would result, although this may vary depending on a number of factors, including the metabolic state of the organism and the fat content of the meat. Doyle and Schoeni (1984) reported D values in ground beef (17–20% fat) between 54.4 and 64.3°C of 39.8 to 0.12 min with a z value of 4.1°C (Table 4.21) . Orta-Ramirez *et al.* (1997) reported D values at a number of temperatures, including a D value of 0.12 min at 68°C. Ahmed *et al.* (1995) investigated the effect of fat content on the thermal destruction of *E. coli* O157:H7 in ground beef and reported a large difference in beef containing 7 and 20% fat. The D value at 50°C with 7% fat was 55.34 min whereas in the presence of 20% fat this increased to nearly double that value (92.67 min). Line *et al.* (1991) also reported that resistance to heat was greater in fatty beef than in lean beef. Jackson *et al.* (1996) demonstrated that in contaminated ground beef patties *E. coli* O157:H7 was

Table 4.21 Heat resistance of *E. coli* O157 in ground beef with different fat contents

Temperature (°C)	*D* value (min)	*z* value (°C)	Fat content (%)	Inoculated product	Reference
54.4	39.8	4.1	17–20	Ground beef	Doyle and Schoeni, 1984
57.2	4.5				
58.9	1.17				
60	0.75				
62.8	0.4				
64.3	0.16				
51.7	78.2–80.1*	7.8–8.3*	2	Ground beef	Line *et al.*, 1991
51.7	115.5–121.0*	8.4	30.5		
57.2	4.0–4.1*	7.8–8.3*	2		
57.2	5.3–7.4*	8.4	30.5		
62.8	0.22–0.30*	7.8–8.3*	2		
62.8	0.47	8.4	30.5		

* Variable *D* value because of different recovery rates on isolation media used.

reduced by 3.0–6.4 log cfu/g at 68.3°C in less than 5 s. The greatest heat resistance was observed in burgers where growth of the organism could not occur. The greatest sensitivity to heat was observed in burgers where active growth had occurred. This is consistent with other studies, which have found increased heat sensitivity when exposed to elevated temperatures during active growth, with increased resistance being conferred when in non-growth phases, such as after entering the stationary phase and during frozen/chill storage.

As achieving the correct lethal effect on any contaminating pathogens is so critical, consideration needs to be given to the following key aspects of cooking instruction formulation:

- cooking method
- variation in heating profile in domestic appliances
- formulation of the product, i.e. fat content
- thickness of the product
- number of products cooked at one time
- preheating of the heating source
- distance from the heating source
- temperature of the product prior to cooking
- length of time of cooking
- frequency of turning the product during cooking.

Such factors are more important for thick, frozen beefburgers where it is known that when the thickness exceeds approximately 20 mm the ability to fully cook the product throughout is compromised by burning the outside of the burger. Cooking instructions need to be validated taking account of the variables described above and appropriately detailed guidance should be provided on the pack. Beefburger packs may therefore need to refer to defrosting frozen burgers for periods of time before cooking as this helps thick burgers to cook effectively without burning. They may also need to give guidance on the actual distance away from the heat source and the turning frequency required to achieve the appropriate cook, together with the preheating required of the heat source. It is important to remember that a cooking instruction cannot cater for all eventualities and therefore it is possible that some consumers may not achieve the correct conditions for cooking the product effectively. Even under standardized conditions it is not possible to obtain reproducible results for applying a cook to a beefburger (Liu and Berry, 1996). Because of this it is important to provide simple forms of additional guidance to allow the consumer to make a judgement on whether the product is cooked effectively or not. The Chief Medical Health Officer of the Department of Health in the UK issued guidance using the description 'cook until the juices run clear and there are no pink bits inside' (Anon., 1993c). This is now present on most beefburger products sold in the UK. However, it is important to note that some research indicates that under certain circumstances raw meat products may loose their red coloration at temperatures that may be below those required to ensure full destruction of high numbers of *E. coli* O157. This is why it is also important to support the previous advice with further recommendations to ensure that the product is cooked throughout and the centre is 'piping hot'. It may also be prudent to encourage consumers to use more precise temperature monitoring of cooking efficacy, such as thermometers, but modifying consumer behaviour to adopt such approaches is likely to take a long time. In addition to the hazard presented to consumers in undercooked products such as beefburgers it should also be remembered that the product itself, together with minced and other forms of raw meat, represents cross-contamination hazards to the consumer's kitchen environment. The UK Advisory Committee on the Microbiological Safety of Foods (ACMSF) recommended that UK manufacturers of minced beef products adopt the US approach for providing food safety advice on products, including information on how to avoid cross-contamination from such products (Advisory Committee on the Microbiological Safety of Food, 1995). While this is mandatory in the USA, in the UK some retailers have placed such food safety tips on their products on a voluntary basis.

In the catering environment or in fast food restaurants the consumer has no option but to assume the beefburger has been cooked effectively and so the restaurant has complete responsibility for ensuring this is carried out properly. Again, the key to controlling the hazard is to have established validated procedures for ensuring that all products, when cooked on the heating equipment provided, achieve the correct internal temperature when cooked according to the prescribed method. Operatives need to be fully trained to ensure that they follow the prescribed procedure and must be aware of the serious consequences of not following the guidance provided. The cooking time and temperature must take account of all of the variables previously described that affect the achievement of the correct internal temperature and it is important that such procedures are supported by checks conducted at the start of the day and repeated at intervals throughout the day. These should include checks on the internal temperatures of products using sterilized temperature probes. Naturally, if the cooking process used do not allow achievement of pathogen destruction temperatures, as appears to have been the case in the outbreaks in the USA associated with undercooked beefburgers from a fast food outlet, then following the correct procedures will be of little consequence. In such situations, it is clear that an understanding of the basic hazard and its control may allow operatives to detect problem situations leading to undercooked burgers and alert management about the need to alter cooking regimes, e.g. times or temperatures. However, it is important to note that given the practice in the USA for consumers to request burgers cooked rare, it is equally important to ensure consumers are themselves educated in the significant difference in the risk associated with consuming beefburgers or beef steaks rare. Such information dissemination can only be achieved by concerted public education programmes about these risks and the need to ensure that a full cook is given to comminuted and reformed meat products.

A number of surveys of raw meat and minced meat products have been reported and although incidence does vary in these reports it is clear that both VTEC and *E. coli* O157 do occur in these products. Smith *et al.* (1991) examined retail chickens and sausages for VTEC using DNA probes for the VTEC genes on isolated colonies. In a total of 184 raw sausage samples, 25% hybridized with the probe although none of the chickens (112 samples from 71 chickens) were found to be contaminated. None of the isolates were *E. coli* O157:H7, although the VTEC isolated included eight different O serogroups. Doyle and Schoeni (1987) reported detection of *E. coli* O157:H7 in retail fresh meat samples: 3.7% in beef (6/164), 1.5% in pork (4/264), 1.5% in poultry (4/263) and 2% in lamb (4/205). Padhye and Doyle (1991) surveyed fresh ground beef in retail stores and detected 2.8% (3/107) contaminated with *E. coli* O157:H7 with levels of contamination ranging from 0.4 to 1.5 cfu per gram. In a large survey in The Netherlands, Heuvelink *et al.* (1996) found 0.3% (2/770) of samples of minced mixed beef and pork to be contaminated with *E. coli* O157 although no samples of raw minced beef (0/1000), raw minced pork (0/260) or poultry products (0/300) were contaminated as individual species of meat. End product testing of meat and meat products has become a routine part of the meat inspection procedures adopted by the USDA after the large outbreaks associated with ground beef. This has been introduced alongside implementation of HACCP at the processing plants and the associated inspection carried out by USDA inspectors at processing facilities. However, testing finished products for the presence of an enteric pathogen that may be present in such raw products, even when operating good conditions of hygiene, has been criticized in the USA by industry representatives. The testing programmes has resulted in costly recalls of raw meat products from the American market. However, it is clear that with the increasing number of incidents of *E. coli* outbreaks implicating raw meat products, large recalls of implicated products (Ault and Morris, 1997) will be commonplace in what the legislators believe is the interest of the public.

The move to animal and carcass decontamination techniques is likely to be furthered significantly by such high demands on the raw meat industry.

READY-TO-EAT SALADS AND VEGETABLES AND SPROUTED SALAD VEGETABLES

In recent years there has been a significant increase in the sale of prepared salads and vegetables as part of the general move to more convenience ready-to-eat and 'healthy' foods. The products are washed, prepared and often sold prepackaged with a limited shelf life of 4 to 6 days under refrigerated storage (<8°C). Products usually include combinations of lettuce, cabbage, onions, carrots, beansprouts, salad cress and other fresh produce. Some products are retailed with additional cooked or cold mixed sachet sauces or dressings for use with the salad. All products are sold as ready-to-eat foods with the customer not being required to wash the product. Salad products have a history of being associated with food poisoning outbreaks (Beuchat, 1996) although these have primarily been associated with enteric pathogens other than *E. coli* O157. Recently, however, sprouting vegetable salads such as alfalfa and radish sprouts have been implicated in outbreaks of infection caused by *E. coli* O157 (Como-Sabetti *et al.*, 1997). Sprouting salad vegetables such as beansprouts are a component of many ready-to-eat salads but may also be sold separately as prepackaged products. Some sprouting salad vegetables may be sprouted and grown in punnets and sold without harvesting, as is the case with mustard and cress and salad cress, which are retailed as entire plants with soil, roots, stem and leaves.

Description of process

Prepared salads and vegetables

The production of ready-to-eat salads and vegetables varies significantly depending on the product and the degree of sophistication in the manufacturing plant. The process starts with crop harvesting, transport and cooling of the salad produce prior to delivery to the processing unit where preparation of the raw salad and vegetable ingredients occurs (Figure 4.7). For example, a raw material such as a lettuce or cabbage would usually be prepared by removing the core and outer leaves while other raw materials such as carrots or onions may be peeled using automated peeling machines. After primary preparation, the majority of the produce is washed in chlorinated water and then excess water is removed by gentle spinning or draining. The various produce types are then cut and assembled into packs prior to sealing and gas flushing of the pack.

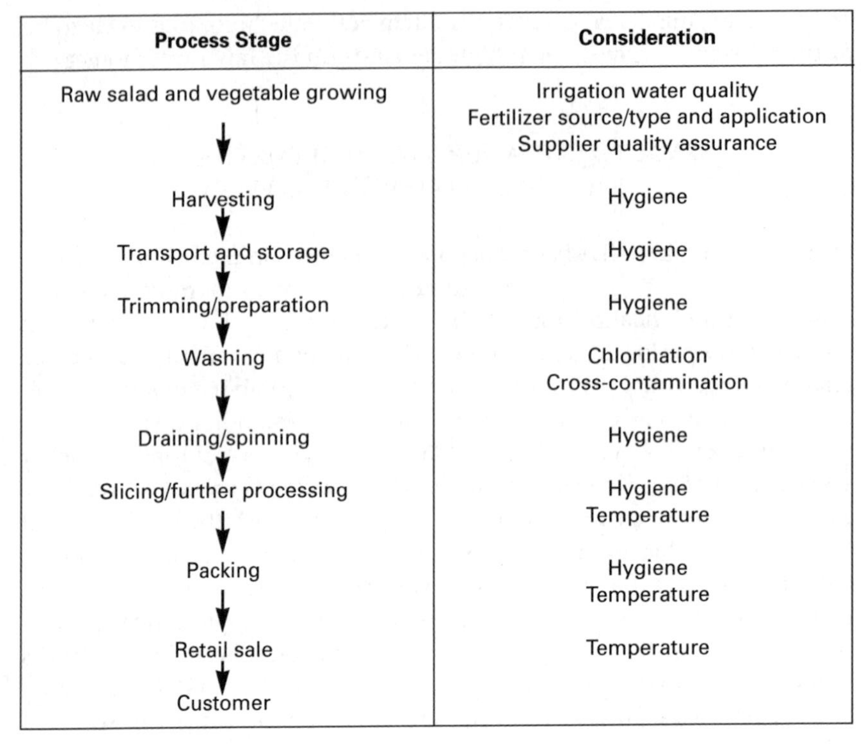

Figure 4.7 Process flow diagram and technical considerations for a typical prepared ready-to-eat salad.

Sprouting salad vegetables

Sprouting salad vegetables such as beansprouts and alfalfa are produced in a completely different manner with the product being grown from seed in the producing unit (Figure 4.8). The harvested seed, mung bean in the case of beansprouts, is rehydrated and then placed in germination rooms where the temperature and humidity are carefully controlled. The temperature is usually kept above 20°C and the seeds receive a regular supply of water, by soaking or spraying, to facilitate germination and growth. Once germinated, the seeds grow rapidly and within several days are usually large enough to harvest. Many sprouted vegetables are moved out of the germination rooms after germination to allow further growth of the plant and formation of appropriate sized foliage. This is usually done in enclosed areas such as greenhouses to ensure that light, temperature and humidity can be controlled. Sprouting vegetables grown in pots or punnets for sale as an intact unit to the consumer are usually grown in peat although the nature of the soil may vary. After collection, sprouted vegetables such as beansprouts are usually washed in chlorinated water and then packed and chilled for retail sale.

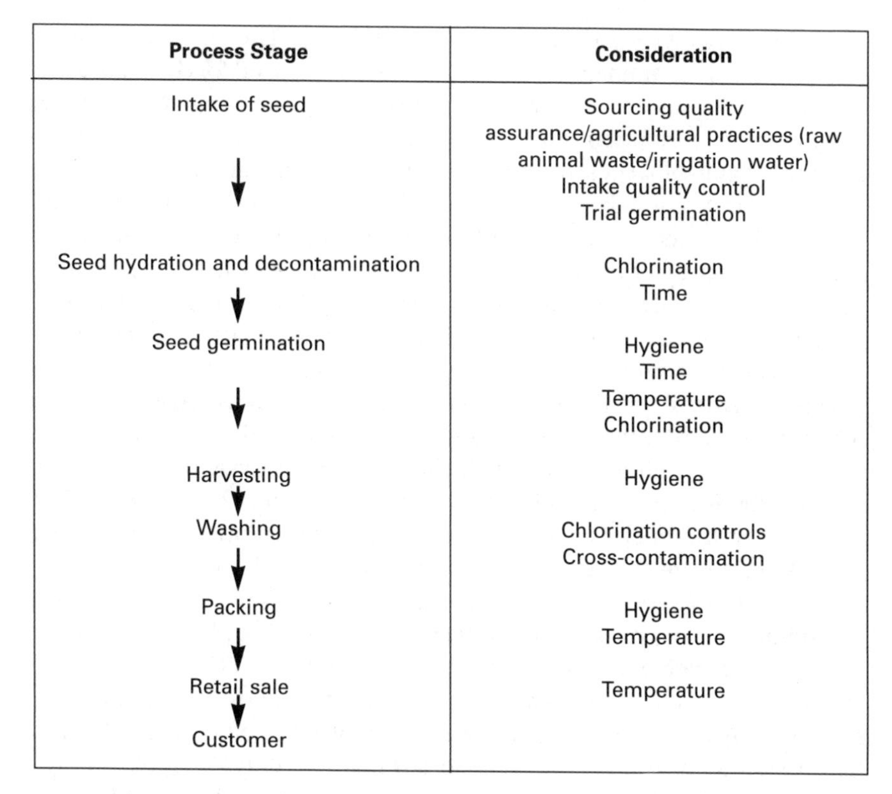

Process Stage	Consideration
Intake of seed	Sourcing quality assurance/agricultural practices (raw animal waste/irrigation water) Intake quality control Trial germination
Seed hydration and decontamination	Chlorination Time
Seed germination	Hygiene Time Temperature Chlorination
Harvesting	Hygiene
Washing	Chlorination controls Cross-contamination
Packing	Hygiene Temperature
Retail sale	Temperature
Customer	

Figure 4.8 Process flow diagram and technical considerations for a typical sprouting vegetable.

Prepared salads and vegetables (including sprouting vegetables) usually have a short shelf life of 4–6 days, which is dictated more by the significant visual and organoleptic deterioration that occurs with these products rather than the potential for growth of contaminating pathogens. In some countries the products may be sold at ambient temperatures either in prepacks or on market stalls as loose commodities where the shelf life is further restricted by the visual appearance of the product, which generally deteriorates in 1 to 2 days.

Raw material issues and control

Prepared salads and vegetables

Prepared salads and vegetables are simply washed and prepared raw agricultural crops. Naturally, the practices and standards in operation in the field will have a marked effect on the quality and microbial contamination load on the raw material itself. It is critical to ensure that standards of good agricultural practice are maintained at all times. The produce is most vulnerable to contamination during growing. Fields used for the cultivation of salad vegetables may receive natural fertilizer applications at some stage during the crop rotation. As *E. coli* O157 and other VTEC may be contaminants of animal faeces, the application of untreated animal waste could present significant hazards to the fresh produce on germination and growth. Best practice would avoid the use of animal waste completely for the cultivation of crops intended for use in ready-to-eat salad products. However, if they are used the risk of contamination by pathogens can be reduced by application of manures that have been subjected to composting processes in which high temperatures develop in the compost, sufficient to reduce or destroy contaminating pathogens. Also, applying manure several months before planting crops will usually result in a reduction in pathogen numbers although the extent of survival may be dependent on the level of contaminants in the faecal material and the level of exposure of the organism to adverse weather conditions such as drying and UV light from the sun. Little data have been published on the potential survival of *E. coli* O157 in animal slurries or manure after application to land but it is of considerable concern that as farming practices result in an increased incidence of *E. coli* O157 and other VTEC in slurries and manure then such contaminants spread onto the land may increase the burden with which the natural processes involved in pathogen reduction have to contend. Maule (1997a) reported that *E. coli* O157 could survive for several weeks in cattle faeces and even longer in soil, although levels decreased steadily with time (see Table 4.11). Nevertheless, with a common sense approach to the use of natural fertilizer it is possible to minimize the hazard presented from this source to the salad vegetable during growing. It is, however, apparent that outbreaks of foodborne illness associated with the application of animal slurries do occur, as was seen in the large outbreak in Germany caused by Vero cytotoxin-producing *Citrobacter freundii*. This was likely to have been caused by the consumption of green butter, where the butter contained organic parsley, for which pig manure was used as a fertilizer (Tschäpe *et al.*, 1995). Mermin *et al.* (1997) reported an outbreak of *E. coli* O157:H7 due to the consumption of a mixture of different types of baby lettuce. Cattle were reported to be found in areas next to the lettuce growing and processing areas. It is essential if using natural fertilizer to ensure that raw animal waste is not applied to land after crops have been planted and especially not to exposed crops.

As well as limiting the potential for contamination of crops by management of the farming practices involved in the use of natural fertilizers for the production of the crops intended for these products, it is also important to ensure that the water used for irrigation of these crops is equally well controlled. Contamination of fresh produce may occur due to the use of contaminated irrigation water. The application of irrigation water to growing produce is often necessary to meet the water requirements of the crops. However, the irrigation water may be applied by aerial spray or by ground irrigation channels. Much of the salad produce grown in the UK is grown in enclosed areas and the quality of the irrigation water can be maintained to drinking water standards, although untreated borehole water may also be in use. In poorly developed countries, however, such standards are not necessarily possible and the only form of irrigation water may be local rivers or streams. As the rivers may also be the only means of disposal of sewage, it is obvious how hazardous situations may arise whereby faecal contaminants gain access to growing produce. Hazards are also presented by field workers in situations where inadequate sanitation facilities are supplied and where little or no education in the need for basic standards of hygiene is given. It is essential that field workers understand their role in controlling the safety of these products, even if such training involves little more than a basic discussion before first working in the field. Naturally the provision of field-based toilets is essential as it is of little help to provide people with the knowledge of how to maintain basic hygiene standards without providing them with the facilities to do so.

To ensure the safety of their finished products, manufacturers of prepared salads and vegetables should, wherever possible, have effective supplier quality assurance programmes that include the conditions under which the raw materials are grown and harvested, and also encompass the basic practices and hygiene standards required for preventing contamination from arising. Products grown under controlled conditions without the application of natural fertilizer and using a potable water supply provide the best opportunity for minimizing the potential for faecal contamination of the raw material. Irrespective of this, it is essential to have an understanding of the hazards and how they are controlled in the following areas:

- application of wastes to land (type and time)
- application of irrigation water to land (source and treatment)
- hygiene of workers involved in harvesting of crops.

Sprouted salad vegetables

For sprouted salad vegetables, such as bean sprouts, alfalfa and similar products, the producer usually purchases the seed in bulk supplies to last for several months. Since much of the enteric pathogen contamination of sprouting vegetables arises in the raw seed itself and any contaminants are then given ample opportunity to increase in number during the germination and growth stages of the sprouting process, it is absolutely essential to ensure that consignments of seed are free from enteric contaminants including *E. coli* O157 and other VTEC or at least that their incidence is minimized. Ideally, the supplier of the seed should be known and visited to assess and ensure the application of good agricultural practices in the preparation of the seed. However, seed is often purchased on the open market from poorly developed countries and the seed origin may be difficult to trace; consequently, assessing the level of agricultural control may not be possible. Wherever possible, seed should be obtained from reputable suppliers with whom purchase agreements can be made, including the implementation of an effective supplier quality assurance programme. Seeds should be subjected, by the producer, to intake testing prior to use, including monitoring for indicator microorganisms such as *E. coli* generally together with specific contaminating pathogens such as *Salmonella*. The usefulness of testing raw material seed for *E. coli* O157 when the incidence of the pathogen is likely to be exceptionally low is limited but the use of composite sample testing of a large number of small samples from all seed bags in the batch delivery can increase the probability of detecting low numbers of contaminants distributed heterogeneously in consignments. Such testing can be supplemented with microbiological assessments after trial germination from batches of seed. By germinating samples of seeds from a representative sample of bags under conditions that the normal seed would be subject to it is sometimes possible to identify whether low level contaminants may compromise the finished product. Trial germination together with intake testing can be a useful means for selecting superior quality seed but, of course, it cannot guarantee the absence of pathogens, which may be present in discrete pockets, and such testing must be seen as only a component part of the supplier quality assurance programme.

Process issues and control

Salad and salad vegetables

The primary area for contamination of salad and vegetables is in the field but once received by the producer they need to be handled appropriately to prevent the introduction of new contaminants or the spread of contamination. Salad and vegetables are usually trimmed and prepared and then washed in chlorinated water. Standards of hygiene need to be maintained to prevent contamination of the salads and vegetables by food handlers. Failures in this area have resulted in documented outbreaks of foodborne illness in the past (Davis *et al.*, 1988). Provided basic good practices are in operation, food poisoning outbreaks from these products can be minimized. The washing stage in the process is critical to preventing the spread of contamination. If inappropriately controlled, washing procedures have the potential to spread contamination throughout entire batches of product. Chlorination of produce washing water needs to be properly controlled to achieve the desired result, which is both to remove physical debris such as soil or dirt and to reduce levels of microbiological contamination while preventing such contamination passing to other products during the washing process. Some products may lose colour in chlorine therefore care must be exercised in the selection of suitable processes. Chlorine levels must be maintained with sufficient free chlorine present continually to allow the treatment to be effective. For the best effect on reduction of microbial load the pH of the chlorine treatment water should be maintained between pH 6 and pH 7. Product should be submerged, agitated and given sufficient contact time in the treatment system. Levels in excess of 100 ppm free chlorine are usually applied for chlorine washing of salad vegetables although levels up to 200 ppm have also been advocated (Beuchat, 1996). Such levels do not usually leave taint on these products as the chlorine is readily inactivated by the organic material present. In fact, it is for this reason that chlorine washing is not especially effective at reducing microbial contamination when compared to water washing. Levels used usually achieve a reduction in excess of two orders of magnitude but the significance of the presence of chlorine is to provide conditions under which cross-contamination to other products in the wash water is prevented. Surface disinfection of produce has been reviewed (Beuchat, 1992) and the use of chlorine at levels of 100 ppm (free chlorine) has been shown to reduce the levels of general microflora on lettuce by approximately 2 log units in comparison to the 1 log unit decrease achieved by washing in tap water (Adams *et al.*, 1989). Adams *et al.*, (1989) also demonstrated that increasing the washing time in chlorinated water (100 ppm free chlorine) from 5 to 30 min had little effect on reducing microbial numbers. Decreasing pH from 8.8 to 4.0 had a marginally improved effect on reducing numbers (Table 4.22), although at such low pH values the chlorine is likely to be unstable. The failure to achieve a significant reduction in microbial numbers on the surface of vegetables was attributed to the protection afforded to the microbial population by the hydrophobic pockets and folds in leaf surfaces. Chlorine washing systems may be manual using dip tanks where the cut vegetable is immersed in perforated trays and held for a defined period of time. Alternatively, they may be continuous flow systems where the products pass down chutes in a flow of chlorinated water, tumbled by paddles or agitated by forced air in a vessel. No

Table 4.22 Effect of washing in chlorinated water (100 ppm free chlorine) at different pH values on the microbial load of lettuce (adapted from Adams *et al.*, 1989)

pH	log cfu per gram
Water control	6.59
8.8	6.04
7	5.79
6	5.97
5	5.84
4	5.8

matter what the system in operation, the critical area for control is the maintenance of high free chlorine levels, which should be checked at regular intervals; automatic dosing systems being the preferable option. It is important to establish clear limits for the level of free chlorine that must be present and the residence time of the material. Care also needs to be taken when washing vegetables to ensure that the systems are not overloaded with produce such that some of the vegetables do not get immersed. This can occur in manufacturing units under pressure for high production throughput where washing residence times may be reduced or systems are overloaded.

Washed products may be manually chopped or chopped using automatic equipment. In addition, packing may occur via automatic depositors or manually. Most processes incorporate a high degree of manual preparation and assembly interspersed with automated machinery. Although not likely to be a source of *E. coli* O157 contamination itself, the environment and equipment need to be maintained to high standards of hygiene to prevent any microbial contamination on the equipment from proliferating and becoming sources of cross-contamination. Cleaning and sanitization routines for equipment and surfaces should be conducted frequently, with cleaning efficacy monitored using conventional microbiological hygiene assessment or rapid ATP bioluminescence hygiene monitoring systems. Environmental hygiene should also be monitored using microbial indicators of hygiene such as *E. coli* and Enterobacteriaceae.

Sprouted vegetables

Sprouted vegetables are produced by soaking seeds in water for short periods of time to allow rehydration. It is recognized that many seeds can

Table 4.23 Effect of chlorine soaking on the survival of *Salmonella* species in alfalfa seeds (adapted from Jaquette *et al.*, 1996)

Chlorine concentration (ppm)	Time (min)	cfu per gram of seed*
0	5	248
	10	387
100	5	128
	10	197
1010	5	8
	10	37

*: initial contamination level of seed: 339 cfu/g

tolerate quite high levels of chlorine during the seed soaking stage and many processors use this feature in attempts to decontaminate seed surfaces prior to full germination. Levels of chlorine usually exceed 100 ppm and although contamination is reduced it cannot be completely eliminated as the contaminants may reside within the seed itself or be protected from the chlorine.

Jaquette *et al.* (1996) studied the effect of chlorine soaking of seeds on the survival of *Salmonella* spp. and although a high reduction was reported, microorganisms could remain in sufficient numbers to then grow during the germination stage of the process (Table 4.23). It is therefore useful to incorporate chlorine in the soaking stage in any sprouting seed process as this can reduce bacterial contamination levels, but it must be remembered that this will not make a contaminated seed batch safe.

After soaking, the seeds are usually transferred to germination rooms where they are kept under controlled conditions of humidity and temperature to encourage the germination process. Beansprouts are usually germinated with the addition of water and are not exposed to soil. Other products, however, may be grown on peat or other soils and this introduces further opportunities for contamination. Germination of seeds at elevated temperatures with high humidity provides the ideal conditions for plant growth but these conditions are also ideally suited to the growth of enteric pathogens. Some processors apply low levels (<5ppm) of chlorine to the water in an attempt to control contaminating organisms but this is unlikely to have a significant effect on the organism as the chlorine will be readily inactivated on the plant tissue and peat, if present. Levels of microorganisms can increase significantly during this stage. Hara-Kudo *et al.* (1997) studied the growth of *E. coli* O157:H7 (which included a strain isolated from the Sakai City outbreak) during the sprouting of radish seed and found that after being soaked in 4 ml water containing log 4 and log 4.1 cfu/4 ml for 8 h, the organism increased to log 8.6 and 9.8 cfu/50 sprouts after a 32 h germination period. They also demonstrated that after soaking seeds in water contaminated with log 3.0 and log 3.1 cfu/ml of *E. coli* O157:H7 and then germinating using a daily supply of fresh water, the levels in the edible parts of the radish sprouts i.e. cotyledons and hypocotyl were log 7.3 and 7.8 respectively. Furthermore, after submerging the roots of radish sprouts in water containing varying levels of *E. coli* O157:H7 (10^4–10^7 cfu/ml) for 18 h, the organism could be detected in a high proportion of cotyledons and hypocotyls (ranging from 6.7% to 100%). The potential for excessive growth of *E. coli* O157 or other VTEC, if present, in the seed or water supply is clearly of significant concern. Once present there is very little that can be done to prevent the organism from growing.

After germination some products are transferred to greenhouse units to allow further growth. During this stage the products receive further water, which in the UK is often of potable water standards to prevent the introduction of microbial pathogens. In poorly developed countries this may be less well controlled and products may be cultivated without protection from potential pathogen contamination and may be irrigated with water from streams or rivers. Once a sprouting vegetable has exposed surfaces, any contamination is likely to remain on the product and therefore control of post-germination contamination is critical. Products are harvested manually or automatically and may receive a further chlorine wash prior to spinning or draining and packing, depending on the product. The controls already mentioned regarding the control of chlorine washing need to be maintained.

Final product issues and control

After preparation and washing, many salads and vegetables, including sprouting vegetables, are packed into sealed pouch packs, which may be gas flushed. The gas flushing usually has little effect on the microflora in the pack but is present to maintain the organoleptic properties of the product during the shelf life. Many products are sold from chilled display counters and if these are maintained under 5°C no growth of *E. coli* O157 would be expected. Some produce is sold at ambient temperatures and because of the presence of plant cell nutrients released during the cutting and preparation of salads and vegetables, growth of contaminating pathogens could occur. Abdul-Raouf *et al.* (1993) studied the growth and survival of *E. coli* O157:H7 on shredded lettuce, sliced cucumber and shredded carrot when subjected to different temperatures and under different storage conditions (gas mixtures). They found that packing under 3% oxygen and 97% nitrogen had little effect on the populations of the organism. The levels of *E. coli* O157:H7 gradually declined on vegetables stored at temperatures of 5°C but increased on those stored at 12°C for up to 14 days (Table 4.24). At 21°C, levels increased by several orders of magnitude within 3 days but then decreased after 7 days due to the accumulation of acids from the prolific growth of other microorganisms. They also noted that carrot appeared to be extremely inhibitory to *E. coli* O157:H7. It is important to recognize therefore that the final product, if contaminated, must be handled and stored carefully to prevent low levels of contaminants from growing during the shelf life of the product.

Surveys of fresh vegetables and salads in the retail market-place frequently report *E. coli* as contaminants. Occasionally *Salmonella* species are also reported but the presence of *E. coli* O157 in routine surveys is rare (Table 4.25). Clearly, it is not possible to completely exclude the potential for contamination with pathogens like *E. coli* O157 but procedures in place from the field to the finished product must attempt to minimize the chances of this occurring. Underpinning all these procedures should be a competent hazard analysis of the specific process. An excellent example of a relevant, generalized HACCP has been developed by the International

Table 4.24 Growth of *E. coli* O157:H7 (log cfu/ml) on prepacked fresh prepared vegetables under different storage temperatures (initial pack atmosphere was air) (adapted from Abdul-Raouf *et al.*, 1993)

Temperature		Days				
		0	3	7	10	14
5°C	Shredded lettuce	5.34	5.06	5.38	4.91	4.23
	Sliced cucumber	5.1	4.6	4.27	3.43	NT
	Shredded carrot	5.31	4.55	+	+	+
12°C	Shredded lettuce	5.34	6.85	7.29	7.48	7.52
	Sliced cucumber	5.05	6.08	5.84	5.74	NT
	Shredded carrot	5.31	7.1	6.78	6.61	6.3
21°C	Shredded lettuce	5.34	8.47	8.83	NT	NT
	Sliced cucumber	5.05	7.26	4.64	NT	NT
	Shredded carrot	5.31	7.36	6.03	NT	NT

+: positive by enrichment.
NT: not tested.

Table 4.25 Surveys of the incidence of *E. coli* contamination in vegetables

Product	Organism	Results	Number of samples	Reference
Cabbage	*E. coli* O157:H7	1 + ve	4	Beuchat, 1996
Celery	*E. coli* O157:H7	6 + ve	34	Beuchat, 1996
Cilantro	*E. coli* O157:H7	8 + ve	41	Beuchat, 1996
Coriander	*E. coli* O157:H7	2 + ve	10	Beuchat, 1996
Celery	*E. coli*	10 samples > 10^3 per 100 g	26	Ruiz *et al.*, 1987
Cabbage	*E. coli*	5 samples > 10^3 per 100 g	41	Ruiz *et al.*, 1987
Lettuce	*E. coli*	31 samples > 10^3 per 100 g	80	Ruiz *et al.*, 1987
Parsley	*E. coli*	13 samples > 10^3 per 100 g	23	Ruiz *et al.*, 1987
Miscellaneous salad vegetables	*E. coli* O157	0	63	Lin *et al.*, 1996
Miscellaneous salad vegetables	*E. coli*	8 + ve	63	Lin *et al.*, 1996

[a] 29/112 colony isolates were identified as *Escherichia*.
[b] 48/97 colony isolates were identified as *Escherichia*.

Fresh Produce Association (IFPA) and this has been summarized by Beuchat (1996). Any processors of these types of products are strongly urged to assess their own processes using the guidance and suggested approaches available.

GENERIC CONTROL OF *E. COLI* O157 AND OTHER VTEC

Raw material identified as a potential hazard

If the raw material is likely to be contaminated with *E. coli* (generic, O157 or VTEC) and no subsequent process exists for reducing it to an acceptable level, e.g. cooking, then control of the raw material is absolutely essential.

In most cases the raw material is not usually under the direct control of the product manufacturer, as is often case with raw milk for raw-milk cheeses, raw meat for salami, apples for apple juice and so on. In such circumstances, it is important to operate a raw material supplier quality assurance programme. The scope of the supplier assurance programme should be as detailed as necessary, tracing the raw material back to source and assuring the adequate control of relevant hazards that may be introduced during all stages prior to receipt. It is recognized that in some circumstances it is difficult to achieve traceability to source and therefore direct audit of practices may not be possible, but major raw materials, such as milk, meat, seeds, etc., should always be supplier quality assured and such programmes should incorporate as many of the following points as possible.

1. Detailed understanding by the raw material user of the production process of the raw material and knowledge of the critical control points or those stages in the process influencing the control of VTEC.

As the microbial safety of the raw material is likely to be the most important issue in the entire production process, a clear understanding of the process involved in the production of the raw material is critical to the ultimate safety of the final product. This knowledge is important in allowing a manufacturer to determine those suppliers who are able to provide a better quality and/or safer raw material supply. Understanding the raw material production process and the hazards likely to arise during that process can themselves form the basis of a food safety audit of the supplier as part of the supplier quality assurance programme and also help to target any supporting microbiological testing of incoming materials to components particularly vulnerable to contamination. Manufacturers should expect as a minimum that their raw material supplier is operating a hazard analysis approach to the identification and control of hazards, although this is not always evident in many smaller raw material suppliers, particularly in the primary agricultural sector. Nevertheless in a choice between a raw material supplier operating an effective HACCP approach and an alternative who has little systematic control of hazards, the decision to purchase raw material must clearly be heavily weighted to the former site.

2. Audit of the raw material supplier to review process control

Clearly, for critical raw materials it is essential that a supplier audit programme is established to review the control of the hazard in question, e.g. *E. coli* O157 or VTEC. Such audits need to be carried out at regular intervals and must focus on aspects impacting on the control of the specific hazard. Knowledge gained from the process information and hazard analysis can be used to focus any audit to ensure that all relevant critical factors associated with product safety are reviewed during an audit. In many cases such audits may be conducted by third parties but they must have a defined scope and objective, and both parties must be fully committed to implementing action points arising from such reviews. Manufacturers of raw fermented meats, for example, should conduct regular inspections of the raw material meat suppliers, usually abattoirs and boning premises, to focus on hygienic processing. A large and beneficial difference can be made to the incoming incidence of *E. coli* by selecting raw material suppliers operating high standards of hygiene, even though it may be impossible to eliminate the hazard completely. Successful attempts to control *E. coli* at this stage can have a major impact on reducing the level of exposure to the community generally and the levels of exposure to individuals.

3. Raw material verification checks

Reliance must never be placed on microbiological testing of raw materials to control a hazard. However, such testing, if focused appropriately, can assist in monitoring the general level of hygiene in operation at the supply site. Clearly, it is not practical to expect a manufacturer of a salami or raw-milk soft cheese to monitor every carcass of meat or farm delivery of milk for *E. coli* O157. However, it may be useful to monitor the incidence of such pathogens as part of ongoing surveys of critical raw materials. In most cases such low frequency testing for specific pathogens is supplemented with routine intake testing for indicators of hygienic processing, such as *E.*

coli generally, coliforms or Enterobacteriaceae, and monitoring trends in this way can be a more cost-effective mechanism for allowing inadequate control to be highlighted and reviewed with the raw material supplier.

A payment incentive scheme based on consistent delivery of high quality raw material is a useful way of encouraging suppliers to maintain high standards of hygiene in the raw material supply. Such incentive schemes have been used in the past by the UK Milk Marketing Board and are currently in use by many raw-milk cheese manufacturers with their supplying farms. There is no reason to believe that the extension of such schemes to the raw meat industry could not also positively influence the standards of hygienic operation.

4. Agreed specification with the raw material supplier

One of the most important aspects of control of the safety of critical raw materials is an understanding by both parties of the need for the highest standards of quality in the production of the raw material. Such understanding can only come from open discussions between the two parties, with particular emphasis made to the supplier that the raw material is to be used in the manufacture of high risk products. Details of the expectations of both parties are usually best documented in a product specification, signed and agreed by the vendor and purchaser.

5. Conditions of storage and use of the raw material

It is important that conditions of storage of the raw material do not in themselves introduce the hazard or provide conditions where the presence of a low level of contamination could proliferate. As *E. coli* O157 and other VTEC do not grow at temperatures below 5°C, the control of perishable raw materials such as meat and milk during storage should be readily achieved.

Production incorporates processes to reduce the level of contamination or eliminate the hazard

Many products involve the manufacture of food where *E. coli* O157 or VTEC may be present in the raw material but a process is applied to reduce the hazard to an acceptable level or eliminate it completely. In such products the key control must be exerted at the stage where the antimicrobial process is applied. Examples of such stages include:

- heat processing for cooked ready meals, cooked meats, etc.

- pasteurization/UHT processing of milk for cheese, milk, yoghurt and other dairy desserts
- chlorination of vegetables for ready-to-eat prepared vegetables and salads
- fermentation and drying for salamis and raw dry-cured meats
- fermentation and maturing for hard cheeses.

As part of any hazard analysis of the production process key critical control points will be identified which, if kept under proper control and monitored, will maintain the safety of the product. The minimum process requirements to ensure reduction of the hazard to an acceptable level must be understood, validated and applied consistently.

In the example of a product with a cooking stage it is essential that the cooking times and temperatures that must be achieved to ensure destruction of the hazard are clearly defined. For cooked meats this may be 70°C for 2 min or an equivalent time/temperature combination and for pasteurized milk it may be 71.7°C for 15 s. These are usually considered as the heat processes below which the product may be unsafe. Systems must be designed and operated to ensure that it is not possible for products to fail to reach the required temperature at all points within the product. Thus, in most cases, processes are designed with a built-in safety margin set at slightly higher temperatures or for slightly longer periods than the minimum to be achieved. If process control systems are designed around target levels and critical limits as defined by the Codex Alimentarius Commission (Anon., 1996g), then under conditions where the process begins to drop below the target level it is possible to take action to rectify the process and bring it back to or above the target before the critical limit is reached. This avoids the expenditure in time and production costs associated with process failure.

It is important to ensure that any system designed to reduce microbial contamination is based on a sound knowledge of the likely impact of changes in the raw material conditions on the subsequent effect of the process. For example, in the cooking of a raw meat such as chicken strips or a beefburger on a belt oven or grill, the meat is placed on a conveyor belt that passes through the oven or grill at a particular speed and then exiting the cooking process in a high risk environment where it is usually chilled. The process is controlled by ensuring that the temperature of the oven or grill is kept above a certain figure and that the belt speed of the conveyor allows a predetermined residence time under the heat source. However, the efficacy of this process will be dependent on variables such as the temperature of the incoming raw material meat and the size of the portions. If some of the meat is thicker than that used during the derivation of the process times and temperatures, then the existing process may be insufficient to achieve the correct temperatures throughout the meat. Likewise, if the process was established with meat entering the oven or grill at a temperature of 4–5°C but a batch is cooked with an ingoing temperature of 0–1°C, then the process may not achieve the correct cook.

Such problems should not occur with the implementation of correctly determined HACCP-based procedures as raw material temperatures and dimensions would be identified for control and monitoring. It is often in situations of outbreaks that relatively simple issues such as these have been found to have been overlooked, possibly because of commercial pressure to manufacture product to meet tight delivery schedules or as the result of a fundamental ignorance of the importance of raw material and process control for ensuring the safety of the final product.

It is important in examples like that of a cooked meat process to ensure that the process is designed to achieve the correct cook under worst case situations and this should be supported by regular monitoring of products post cooking to ensure correct internal temperatures have been achieved.

In the pasteurization of milk both the process and product temperatures are monitored and so inadequate pasteurization is unlikely to arise. Also, modern pasteurizers can usually be set to automatically divert milk away from finished product if the target or critical temperature is not achieved. The use of manual override switches on pasteurizers needs to be carefully controlled as inappropriate use can lead to improperly pasteurized milk being allowed forward during periods of manual override.

In processes where the lethal effect on *E. coli* O157 and VTEC is less pronounced, but where control is nevertheless important for its contribution to final product safety, it is still essential to employ positive process controls. For example, the chlorination of salads and vegetables for use in ready-to-eat prepared products is not recognized as delivering a large reduction in the levels of contaminating pathogens. However, it does reduce levels of contamination by one or more orders of magnitude (Beuchat, 1996) and, as such, should be effectively controlled to ensure the best results are achieved. The use of common bulk tanks of chlorinated water to wash different materials could itself represent a major opportunity for widespread cross-contamination of different salads and vegetables if chlorine levels are not adequately maintained. In such circumstances the control of chlorine levels is essential and the use of continuous dosing systems are preferable, with frequent monitoring of the levels of free chlorine in solution. For the best effect on reduction of microbial load the pH of the chlorine treatment water should be maintained between pH 6 and pH 7. In addition, the contact time with the product and the nature of any system used to facilitate contact, e.g. submersion and agitation, are important areas that warrant control and monitoring.

Where control is exerted by fermentation and drying it is equally important that the fermentation profiles and drying profiles, that are validated and established to ensure control of the hazard, are actually monitored and reliably adhered to for subsequent batches. It is common with most fermented and dried products like salami or cheeses to control process times, temperatures and humidity where relevant. Such process parameters do not necessarily reflect the factors controlling survival or growth of *E. coli* O157 in the product and specific product and process characteristics may need to be considered. Therefore, while milk may normally be expected to produce an acidified curd after incubation with a starter culture for a defined period of time at an appropriate temperature, it is essential to monitor the activity of the starter culture in the product by way of pH decrease or acidity increase. In this way any factors interfering with the activity of the starter culture will be manifest by changes in the fermentation profile. This is also true of pH decreases and acidity development in salami manufacture. An additional factor in the effective control of VTEC during salami manufacture is that of moisture loss; monitoring moisture loss against a validated profile provides additional assurance of effective process control.

It is most important in the safe manufacture of food to clearly identify what is controlling the hazard, i.e. *E. coli* O157, and also to define what the process limits are that will allow effective control. The position of the salami or raw dried meat in the drying chamber may influence its rate of fermentation or drying. The position of a cooked meat in an oven may influence the achievement of the desired heat process. All such factors must be taken into account in the design and validation of a safe process before any production ever begins.

Product could be recontaminated with *E. coli* O157 or VTEC as a post-processing contaminant

E. coli O157 or other VTEC are not recognized as being common food factory environmental contaminants and therefore opportunities for post-processing contamination should be fairly limited if systems are in place incorporating the basic principles of good manufacturing practice.

The principle hazards from post-process contamination arise in two key areas: from personnel directly handling product and from cross-contamination between raw food materials and the finished product. As *E. coli* O157 is infectious at extremely low levels, the procedures in place must minimize the opportunity for any post-process contamination. Control of environmental and equipment hygiene is primarily of importance if the organism gains access to the post-process manufacturing plant as it then becomes critical that the post-process hygiene practices and cleaning procedures effectively limit any spread of the hazard while removing it from the plant during cleaning.

Raw/cooked separation

Any process in which a raw material is used that may be contaminated with *E. coli* O157 and which incorporates a destruction or reduction stage, resulting in a finished product where the hazard has been eliminated or reduced to an acceptable level, is vulnerable to the hazard of cross-contamination between the raw material and the finished product. It is critical to the safety of the final product that systems are in place to prevent the raw material from coming into contact with the finished product. Cross-contamination can occur in two ways. First, it may result from direct contact of the two materials due to poor segregation of the raw materials from the finished product. For example, in the process of cooking meat ,raw meat is loaded into the oven and, after cooking, the cooked meat may be removed through the same oven door. If the cooked meat is placed next to a batch of raw meat then the consequences are obvious. Alternatively cross-contamination may occur indirectly if cooked meat is placed on the same surfaces used for raw meat and the surface becomes the vehicle of cross-contamination. In many cases it is the operatives themselves that provide most opportunities for cross-contamination when operating practices allow them to handle both raw and cooked meats. Contaminants can be readily transferred by hands, clothing, product contact surfaces or utensils. Clearly, simple rules must exist to prevent the opportunity for cross-contamination of this nature. Factory process flows should be designed to ensure that raw materials arrive at one point and are kept completely separated from finished processed products, preferably by physical barriers. In the case of cooking processes the divide between the low risk side and the high care side is often the oven itself. This may be a double-entry system in which the raw meat is loaded into the oven from one side and, after cooking, the cooked product is unloaded from a door on the other side. The two doors cannot be opened at the same time and the oven is usually built into a dividing wall so that personnel cannot pass between the two areas. It is common practice in large manufacturing units for personnel to be dedicated to either the raw side or the finished product side to prevent opportunities for direct cross-contamination, although in small operations or in retail environments such division is not usually achievable. Prevention of cross-contamination in these latter situations relies on the application of basic principles of good hygienic practice, i.e. using separate utensils and surfaces for raw and cooked products and ensuring personal hygiene practices are maximized by regular cleaning of hands after handling raw products and before handling cooked foods. The same surfaces should not be used for raw and cooked foods; separate, dedicated chopping boards and utensils should be used. Although primarily designed for the control of *Listeria monocytogenes* the principles of high/low risk separation and the procedures that need to be implemented for effective control are detailed in the Chilled Food Association guidelines (Anon., 1997d).

The best way to control enteric pathogen hazards in potential cross-contamination situations is to systematically identify the hazards and required control using a structured hazard analysis for examining all production practices and assessing whether any of these could allow contaminants to pass from the raw material to the finished product.

Personnel handling product

Any product that is extensively handled is undoubtedly exposed to a high risk of bacterial contamination. Clearly, *E. coli* O157 is highly infectious and the significant amount of reported person-to-person spread of the organism during outbreaks demonstrates that transfer from people to foods could also occur. High standards of personnel hygiene are essential to minimize contamination from this source and staff education is vitally important.

It is essential that any person involved in the manufacture or handling of food is fully aware of the hazards that they may contribute to the product by their actions. An understanding of the principles involved in transferring microorganisms from people to the product by inadequate hygiene standards is far more beneficial than any strict work instruction. An educated individual can make an informed decision about practices that are not detailed in work instructions. Uneducated adherence to instructions may allow improper practices to develop and continue until they are identified by trained individuals or until a problem occurs, which is often too late.

It is expected that a clear infectious disease policy will cater for situations in which persons who are obviously ill understand the need to inform their employers and remain away from work. However, it is likely that such persons would not be capable of working in the first instance and therefore the greater hazard exists from those who may be asymptomatic excreters of the organism. It may be better policy for every food handler to assume that they may, from time to time, be carriers of infectious disease agents and therefore they have a clear responsibility to ensure that their personal practices of hygiene control prevent the access of such pathogens into the food chain.

The primary area of cross-contamination from personnel arises from the hands and training should emphasize the hazard presented by poorly cleaned hands. While the use of gloves will undoubtedly reduce the risk of cross-contamination from personnel themselves, it has been the experience of many manufacturers that a glove policy encourages poorer hygiene practices because of the perceived barrier introduced between the operator and the food by the gloves. Use of gloves may be easier to control for employees always working at only one stage in a process, such as in placing meat onto sandwiches, but even in such situations glove washing, changing and disposal procedures must be clearly defined to prevent abuse. The use of gloves in the retail environment, where a person may be serving customers both ready-to-eat foods and raw meats, is likely to be less effective as the need for continued changing of gloves will soon lead some into the temptation of keeping the same gloves on for raw and cooked foods because of time and cost considerations.

Raw products where *E. coli* O157 or VTEC may be present and customer cooking is designed to eliminate the hazard

Raw products in which *E. coli* O157 or VTEC may be present but where the product is designed to be fully cooked by the caterer or consumer do not present a hazard unless they are not cooked appropriately. Under such circumstances product cooking guidelines are critical for conveying the message to the consumer about the need to ensure the product is fully cooked. Cooking guidelines can never cover all eventualities but it is essential that when they are provided they are based on practical trials that show they can achieve the specified cooking temperature and time. Cooking instructions must be validated by the manufacturer or retailer generating them. They must take into account the different types of cooking appliance used in domestic kitchens and should evaluate the effect of distance from the heat source, turning frequency, thawing conditions and product characteristics on the achievement of the final cooking temperature. These considerations have been more fully reviewed in previous sections.

In most cases a cooking instruction for a raw product should be accompanied by a statement that informs the customer of the need to ensure that the product is cooked until it is 'boiling' or 'piping hot' as additional advice to supplement the instruction itself.

Advice for susceptible groups

Some product types are, and are likely to remain, a greater hazard to human health in respect of *E. coli* O157 and VTEC than others, even under the best conditions of manufacturing control. If it is not possible to reduce the hazard to a sufficiently low level then it is important to inform the public, and especially susceptible groups, about the hazards that may be inherent in the products so that they can avoid them.

Little advice has been given to the public in relation to foods to avoid because of the potential for infection with *E. coli* O157 or VTEC. Most emphasis has been placed on the control of the hazards of cross-contamination and undercooking of raw comminuted meat products such as beefburgers. Health departments in the USA have taken the lead on highlighting consumer practices that can control the hazard, with excellent communication programmes reinforcing the message to cook burgers properly and avoid cross-contamination of raw meats to ready-to-eat foods. In the USA, it is now a mandatory requirement to include food safety messages on the packaging of raw meat products, highlighting to consumers that they are handling a product potentially contaminated with pathogenic microorganisms. In the UK, such advice is given in the form of general leaflets on food safety issued by the Ministry of Agriculture, Fisheries and Food (Ministry of Agriculture, Fisheries and Food, 1991) but it is also common practice for individual retailers and manufacturers to provide focused advice on food safety to their customers. Many retailers and manufacturers of burger-type products have also voluntarily reinforced food safety messages by placing additional advice on product packs such as 'cook this meat product until the juices run clear and there are no pink bits inside'. Some retailers have also added basic food safety tips on the food packaging of raw meats and raw meat products.

5

INDUSTRY ACTION AND REACTION

INTRODUCTION

E. coli is one of the most common test organisms specified in microbiological criteria for foods appearing in a variety of regulatory criteria, industry guidelines and purchase specifications. It is frequently used as an indicator of the hygienic processing of foods and it is still used by some as an indicator of the likelihood of the presence of enteric pathogens. However, the usefulness in the latter case is limited as the organism may be present without the concomitant presence of pathogens.

Historically, it has been rare for the food industry to examine foods for the presence of specific types of pathogenic *E. coli* as suitable methods have not been available for general laboratory use. It is only since the recent emergence of *E. coli* O157 and the development of specific methods for detecting this serogroup that food products and associated raw materials have started to be screened for this specific pathogen.

It is questionable whether there are any benefits to be gained from routine testing programmes for *E. coli* O157 in foods as, when present, it is at such a low frequency and usually at such low levels that the chances of detecting it are very small. Nevertheless, routine testing for *E. coli*, supplemented with targeted surveillance of raw materials and processed foods of particularly high concern for the incidence of specific pathogenic groups such as O157, may provide useful verification of process control procedures.

A microbiological criterion consists of statements concerning the microorganism or microbial toxin of concern, the specific food and sample type, the sampling plan to be used, the test method to be used (the method must have been validated for the microorganism or toxin of concern in the food being examined) and the microbiological limit(s) to be applied (the interpretation to be placed on the result and a reaction procedure for those results that are in excess of the upper limit set should also be indicated). There are a variety of texts available that address these areas in detail (National Research Council, 1985; International Commission on Microbiological Specifications for Foods, 1986; Anon.,1996g).

For the food industry, microbiological criteria fall into three categories:

1. Standards: these are microbiological criteria contained in a law. Compliance is mandatory. Examples include most criteria in European Union (EU) Directives and Statutory Instruments of England and Wales. Standards are monitored by enforcement agencies.
2. Guidelines: these are criteria applied at any stage of the food production and distribution system to indicate the microbiological condition of a sample. They are for management information and to assist in the identification of potential problem areas.
3. Specifications; these are microbiological criteria applied to individual raw materials, ingredients or the end product. They are used in purchase agreements.

LEGISLATION AND STANDARDS

The general approach taken to legislation in Europe and North America in the context of food safety is to indicate the clear responsibility of food business proprietors to produce and supply safe and wholesome foods. For instance, in the UK, the Food Safety (General Food Hygiene) Regulations, 1995 (Anon., 1995d) (which implement parts of the EU Directive 93/43/EEC of 14th June 1993 on the hygiene of foodstuffs), Section 4(1) (Anon., 1993b) states:

> 'A proprietor of a food business shall ensure that any of the following operations, namely the preparation, processing, manufacturing, packaging, storing, transportation, distribution, handling and offering for sale or supply of food, are carried out in a hygienic way.'

Further, in Section 4(3):

> 'A proprietor of a food business shall identify any step in the activities of the food business which is critical to ensuring food safety and ensure that adequate safety procedures are identified, implemented, maintained and reviewed on the basis of the following principles:
>
> (a) analysis of the potential food hazards in a food business operation;

(b) identification of the points in those operations where food hazards may occur;
(c) deciding which of the points identified are critical to ensuring food safety ("critical points");
(d) identification and implementation of effective control and monitoring procedures at those critical points; and
(e) review of the analysis of food hazards, the critical points and the control and monitoring procedures periodically, and whenever the food business's operations change.'

Clearly the severe nature of *E. coli* O157 and other VTEC infections make this group of organisms an essential consideration in the hazard analysis of many food business operations. The use of a structured hazard analysis approach to identify means of controlling *E. coli* O157 or VTEC is also likely to generate many of the requisite controls for other pathogenic *E. coli*, but it is important not to overlook the potential hazards presented by the occurrence of other *E. coli* groups.

In addition to the general but important and necessary responsibility imposed by legislation on food business proprietors, other legislation (sometimes referred to as vertical legislation because it deals with a specific food in contrast to horizontal legislation, which applies to generic food production controls, e.g. food hygiene) may also apply, depending on the food type and business. It is, of course, the responsibility of the food business proprietor to know and understand which legislation applies to the business and ensure compliance. Some legislation contains microbiological standards and compliance is compulsory. In addition, some microbiological criteria given in legislation are designated as guidelines, e.g. coliforms, plate incubation at 30°C and plate count (Anon., 1992b).

E. coli is specified in the EU Milk Products Directive (92/46/EC) (Anon., 1992b), implemented in the UK as the Dairy Products (Hygiene) Regulations, 1995. The following standards (cfu/ml or g) are laid down for products on removal from the processing establishment:

(i) cheese made from raw milk or from thermized milk: *E. coli* $m = 10\,000, M = 100\,000, n = 5, c = 2$
(ii) soft cheese made from heat-treated milk: *E. coli* $m = 100, M = 1000, n = 5, c = 2$.

(note: n is the number of sample units comprising the sample, m is the threshold value for the number of *E. coli*, the result is considered satisfactory if the number of *E. coli* in all sample units does not exceed *m*, *M* is the maximum value for the number of *E. coli*, the result is considered unsatisfactory if the number of *E. coli* in one or more samples is *M* or more, *c* is the number of sample units where the *E. coli* count may be between *m* and *M*, the sample still being considered acceptable if the *E. coli* count of the other sample units is *m* or less).

If the standard *M* is exceeded in the case of cheese made from raw milk or thermized milk or soft cheese, testing is required to be carried out to determine whether the *E. coli* strain found is pathogenic and, if so, the relevant products should be withdrawn from the market. No specific mention is made of a requirement to test foods directly for the presence of pathogenic *E. coli* or *E. coli* O157 or other VTEC, although the need to further investigate the nature of any *E. coli* isolated from any samples not conforming to the standards infers that such tests must be made in retrospect. *E. coli* O157 will, of course, be 'captured' in the generic requirement in this milk products legislation, which states that pathogenic microorganisms and toxins from pathogenic micro-organisms must not be present in such quantities as will affect the health of the consumer.

E. coli is also specified in other EU Directives, for example in Council Directive 94/65/EC of 14 December 1994 laying down the requirements for the production and placing on the market of minced meat and meat preparations; microbiological criteria include *E. coli* and minced meat must comply with the criteria; $m = 50$ per gram, $M = 5\times10^2$ per gram, $n = 5$ and $c = 2$ (when a solid medium is used). The microbiological criteria laid down in this legislation are complicated by the fact that the limits for *M* differ according to the test method used (solid or liquid medium) and an additional microbiological criterion is introduced, i.e. the microbic limit value (S). For *E. coli* in minced meat this is set at 5×10^4 per gram. In this Directive and the UK Regulations implementing this Directive (the Minced Meat and Meat Preparations (Hygiene) Regulations, 1995, Statutory Instrument No. 3205), *E. coli* results that are at or above the microbic limit are regarded as toxic or tainted and present an imminent health risk, and relevant products must be withdrawn from the market. For *E. coli* in meat preparations, the criteria applicable are $m = 5\times10^2$ per gram, $M = 5\times10^3$ per gram, $n = 5$ and $c = 2$ (solid medium). No reference is made in this legislation to specific types of *E. coli*. Reasons for the notable disparity in numbers of *E. coli* deemed to be unacceptable for different types of ready to eat dairy products and raw minced meat are far from clear.

Following an outbreak of foodborne illness due to *E. coli* O157:H7 in the USA that was associated with beefburgers, the United States Department of Agriculture Food Safety Inspection Service (USDA-FSIS) commissioned a large-scale investigation into the incidence of pathogens in raw meat and poultry products (Anon., 1994b). This formed the basis of its proposals to set a 'zero-tolerance' (negative result in a specified quantity of meat, e.g. 1 lb) for *E. coli* O157:H7 in raw ground (minced) beef . These proposals were introduced alongside the full-scale implementation of HACCP into slaughterhouses and meat production facilities, and were designed to encourage the industry to adopt improved procedures to reduce contamination in raw agricultural products.

In recent years in the USA the confirmed presence of *E. coli* O157:H7 in raw ground meat has generally led to a 'voluntary' recall by the manufacturer in which contaminated product is withdrawn from retail trade and either destroyed or reprocessed by a means capable of eliminating the hazard. This policy has met with significant criticism as processors unfortunate enough to have had to recall their raw meat products in which *E. coli* O157:H7 (which will inevitably be present at some low incidence) was actually detected have suffered significant financial losses associated with the concomitant withdrawal and destruction or reprocessing of their products.

Raw meat will be contaminated on some occasions with enteric pathogens. While the target of achieving raw meat supplies with little or no such contamination is highly laudable and undoubtedly the imposition of a criterion specifying the absence of *E. coli* O157 has focused processors' attention on the need to minimize its occurrence, the occasional presence of *E. coli* and VTEC clearly remains a possibility. The penalty for its presence in raw meats should, perhaps, not be financial hardship but rather encouragement to invest in systems to reduce incidence further. The ultimate goal must be to eliminate the hazard but the realistic target for raw meat product producers and processors is the minimization of its occurrence.

This common-sense view finds endorsement in a document prepared for the Codex Committee on Food Hygiene by the International Commission on Microbiological Specifications for Foods (International Commission on Microbiological Specifications for Foods, 1996b). The Commission considered it impossible to ensure that animal carcasses are free from EHEC and its presence in raw meats probably reflects its rare incidence in the live animal. In addition, because of the extreme variability of distribution and low frequency of occurrence in meats, the Commission considered that acceptable and practical sampling plans could not be developed for EHEC in raw meat. It also concluded that 'a microbiological criterion cannot presently be elaborated for non-O157 EHEC strains because of the lack of practical methods for the detection of this wider group of pathogens. Although the Commission accepted that criteria could be developed for *E. coli* O157:H7 in foods, because the specific methodology used will not detect other EHEC organisms they felt that this may create a false sense of security as the absence of *E. coli* O157 does not mean that other EHEC will not be present in the foodstuff.

GUIDELINES

Microbiological guidelines used in industry are rarely published as they are generally developed in association with particular processes and products; consequently, a degree of confidentiality applies and industry guidelines remain in-house and self-imposed. Results from routine tests for *E. coli* exceeding the anticipated normal levels for the sample in question should lead to investigations to identify the source and rectify the cause.

General guidelines on the levels and types of microorganism relevant in specified foods produced under good manufacturing practice may be provided by industrial associations for their members, e.g. the Institute of Food Science and Technology (IFST) guidelines on the development and application of microbiological criteria for foods (Anon., 1997f). These include microbiological guidelines for *E. coli* and *E. coli* O157 for a variety of products, including ready-to-eat foods for which good manufacturing condition limits are given as less than 10 per gram and not detected in 25 g, respectively.

In the UK, the Public Health Laboratory Service (PHLS) has published microbiological guidelines (Anon.,1996f) to assist food examiners and enforcement officers in assessing the microbiological quality of foods and to indicate levels of certain types of bacterial contamination considered to be a potential health risk in ready-to-eat foods at the point of sale. Table 5.1 shows how various levels of *E. coli* and *E. coli* O157 are classified for all food categories considered in these guidelines. Although these guidelines have no statutory position, certain actions to be taken in respect of results being found in the 'unsatisfactory' and 'unacceptable/potentially hazardous' categories are discussed. For example, if a result is obtained from a test for *E. coli* O157 classifying a food as 'unacceptable/potentially hazardous' urgent action is recommended to locate the source and identify the risk this presents to

Table 5.1 Guidelines for the microbiological quality of various ready-to-eat foods (Anon., 1996f)

Category	*E. coli* (total)	*E. coli* O157 and other VTEC
Satisfactory	< 20 per gram	Not detected in 25 g
Borderline, limit of acceptability	20–< 100 per gram	
Unsatisfactory	100–< 10^4 per gram	
Unacceptable/potentially hazardous	≥ 10^4 per gram	Present in 25 g

consumers. Results in this category could also form the basis for a prosecution by enforcement officers.

SPECIFICATIONS

Product specifications drawn up between a food manufacturer and a customer (often a retailer) usually include information concerning the physical appearance of the product, physico-chemical characteristics of importance to the safety and/or quality of the product and microbiological parameters relevant to the safety and quality of the product.

There has been a tendency in the recent past (still persisting in some areas of the industry) to establish and maintain a fixed list of microorganisms (potential pathogens as well as non-pathogens and general microbiological tests such as total colony counts and coliforms) that is then included in all product specifications regardless of relevance to the product or the processes by which it was made. For all the reasons discussed, it is generally inappropriate to include *E. coli* O157 in these specifications but it is now increasingly included, because of media attention, and requirements for tests for the organism are often indiscriminately applied to many products and processes where it is of little significance.

If, after consideration of all the issues relevant to determining where and how processes and products need to be monitored, microbiological testing and associated criteria are regarded as useful, then it is important to ensure that only those organisms relevant to the raw materials, processes and finished product are selected (Anon., 1997f). Because very low levels of Vero cytotoxigenic *E. coli* O157 may cause illness, where microbiological specifications are applied, the target level specified for ready-to-eat foods would need to be not detectable in 25 g with an 'unacceptable limit' of detection of the organism in 25 g. Confirmed detection of the organism may lead to removal of products from the distribution and retail system.

MONITORING FOR *E. COLI* AND *E. COLI* O157

The success or otherwise of any systems put in place to control pathogenic microorganisms is usually monitored by testing samples taken from selected points. Where the organism is likely to be present at a low frequency or in very low numbers in the raw material or finished product, as in the case of *E. coli* O157, testing for the organisms may be of limited benefit. Under such circumstances, testing for indicators of contamination such as the broader *E. coli* species may provide better information relating to process control.

Samples for microbiological examination purposes are commonly taken from incoming raw materials, food materials in process, e.g. after washing procedures, cooking procedures or slicing operations, and finished products.

The buying specification for some raw material supplies may include criteria to be met for *E. coli* or *E. coli* O157, e.g. prepared cooked meats for sandwich production. In addition to obtaining a microbiology report from the supplier, incoming batches may be sampled for testing by the sandwich manufacturer. In this case, small portions (10–20 g) may be aseptically removed from several packs, combined and mixed prior to a relevant test portion being used in the laboratory examination. Results from tests for *E. coli* generally provide more useful information and better assurances of hygiene control than tests for *E. coli* O157, although the conditions of a purchase agreement may necessitate testing for the pathogen.

Where in-process samples are required to be taken to monitor the efficacy of production processes, then random single samples (>25 g each) may be taken during the production run and tested individually for indicator microorganisms such as *E. coli*. The results from these tests are assessed against in-house criteria.

Finished product samples are usually taken as complete finished packs from the end of the production line. Single packs may be taken at intervals during the production run. The numbers and frequency of sampling is usually in accordance with a customer's requirements. For monitoring purposes it is often useful to take product samples from the start of the production run as any residual contamination on the equipment may be detected in the first batches of product from the line. In the laboratory, it is important that a representative sample is removed from the pack for testing. For multicomponent products this may involve selecting small quantities from each of the components in similar proportion to their percentage in the product so as to make up a final test portion of 25 g for tests to detect specific organisms at low levels (presence–absence tests) or 10 g for the preparation of a primary dilution for enumeration purposes. Alternatively, the test may be conducted on each individual component or made up from those components deemed to be the highest risk in respect of contamination. Again, the approach taken is usually in accordance with a customer's requirements but for most buying specification purposes the first approach is used. The other options may be applied in problem-solving situations, e.g. if *E. coli* is found in the composite sample tested and a component analysis is warranted.

Environmental samples are commonly taken from food contact surfaces, floor contact items, e.g. wheels and footwear, and air. If meaningful results are to be obtained, then it is important that samples taken are relevant and representative of the area or material targeted. The objective of any sampling programme must be to find any contamination of specific concern so that defined measures can be focused to improve control in areas where the hazard is identified.

Medical swabs are often used for routine environmental sampling where tests are required for monitoring levels of general hygiene indicators such as coliforms, Enterobacteriaceae or *E. coli* on food contact surfaces. In the event, however, that environmental monitoring specifically for *E. coli* O157 is required, e.g. environmental survey work or 'trouble-shooting', then, in order to ensure the best opportunity for detecting the organism if it is present, samples taken should be large in volume or weight or from large areas (using sponge swabs) so as to be more representative of the environmental area and to take account of the likely low level and incidence of the organism.

All sampling procedures, methods, criteria and reaction procedures to be followed in the event of positive results must be clearly indicated in standard operating procedures and the responsibilities of key personnel at all monitoring, reporting and control levels clearly defined. Table 5.2 indicates some approaches to action to be taken in response to results from tests carried out to detect or count *E. coli* O157 in environmental or food samples. These would normally form part of an internal quality system.

Table 5.2 Considerations to be taken into account when finding *E. coli* O157 in product or environmental samples[a]

Consideration	Not detected in any samples	Present in environmental samples	Present in 25 g or greater (ready to eat food)	Present in 25 g to < 100 per gram (raw food)	Present at > 100 per gram (raw food)
No action	Yes				
Environmental monitoring: increase sampling points to identify source		Yes	Yes	Yes	Yes
Cleaning efficacy: check procedures/cleaning records and monitor pre and post cleaning		Yes	Yes	Yes	Yes
Raw material testing: increase testing to identify any contamination			Yes	Yes	Yes
Intermediate product testing: increase testing of in-process material to identify stages of contamination			Yes	Yes	Yes
Finished product testing: increase to identify point of production where contamination is occurring			Yes	Yes	Yes

Table 5.2 Continued

Consideration	Not detected in any samples	Present in environmental samples	Present in 25 g or greater (ready to eat food)	Present in 25 g to < 100 per gram (raw food)	Present at > 100 per gram (raw food)
Review efficacy of process controls: check all process records to ensure current controls have been carried out properly			Yes	Yes	Yes
Stop production: cease production until the problem is identified and resolved			Yes		Yes
Withdrawal of product: consider the need to withdraw products from sale			Yes		Yes

[a] Information given is for guidance only and may not be appropriate for individual circumstances.

In any investigation by public health or enforcement officers of a case of foodborne infection, the ability to demonstrate well-structured and reliably operated procedures targeted to control *E. coli* O157 and other VTEC will prove invaluable.

6

TEST METHODS

CONVENTIONAL METHODS

The key reason for *E. coli* having become such a widely used microorganism in research laboratories is that it grows so readily on simple media and is also easily identified by relatively few biochemical tests. Certainly by the time of the publication of the first edition of Topley and Wilson's *The Principles of Bacteriology and Immunity* (1929) the biochemical characteristics of *Bacterium coli*, as the organism was known at the time, had been extensively examined and the basis established for the now classical IMViC series of tests (Table 6.1) used to differentiate *E. coli* from most of the other members of the Enterobacteriaceae.

Food industry laboratories generally use conventional methods in tests for the detection and identification of *E. coli.* Most often, the methods applied are for enumeration (counting) of *E. coli* in foods because where specifications or standards include *E. coli* it is usually as numbers rather than as presence or absence (detection). Broth methods, in the form of the most probable number technique, used particularly by water industry laboratories, and direct plating methods are both commonly used for enumerating *E. coli*. Figure 6.1 outlines the procedures for both of these approaches, which are now well established in industry, national and

Table 6.1 IMViC reactions of *E. coli* and some other members of the Enterobacteriaceae

Test	*E. coli*	*E. blattae*	*Salmonella* spp.	*Citrobacter freundii*	*Enterobacter cloacae*
Indole	+	−	−	−	−
Methyl red	+	+	+	+	−
Voges-Proskauer	−	−	−	−	+
Citrate utilization	−	26–75% strains +	+	+	+

− = 90–100% of strains are negative; + = 90–100% of strains are positive.

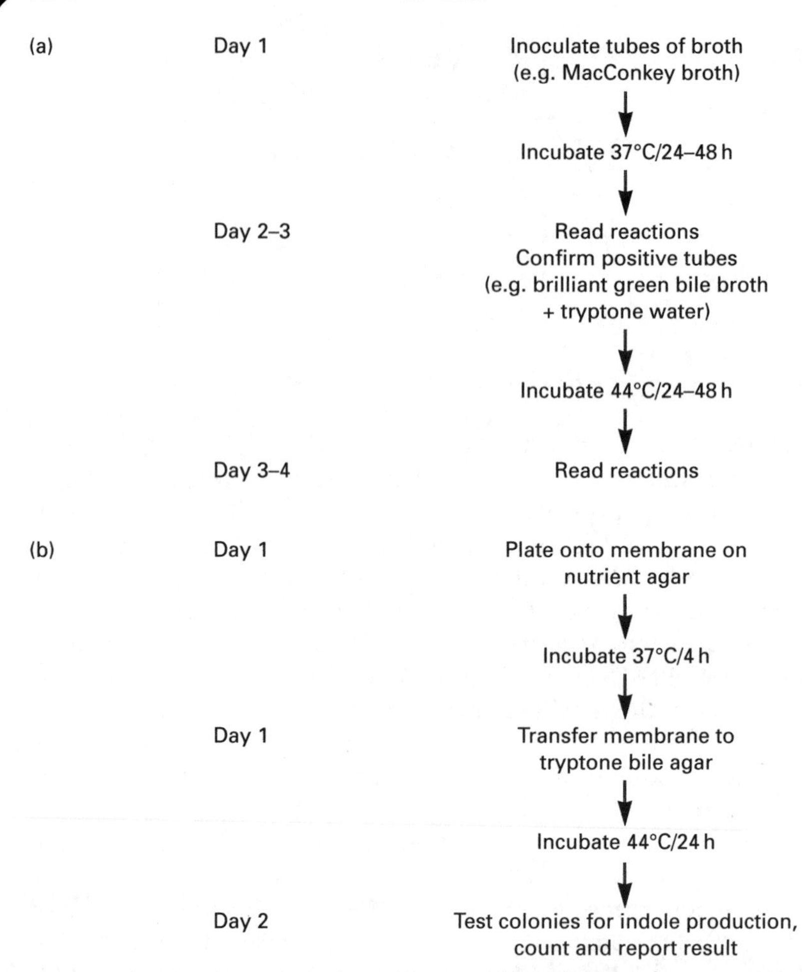

Figure 6.1 General common conventional procedures for the isolation and identification of *E. coli*. (a) broth method (most probable number); (b) membrane method using 1 ml of diluted sample as primary inoculum.

international standards (Anon., 1994c; Anon., 1997h; Roberts *et al.*, 1995; Anon., 1997g). The media used in conventional methods include one or more components targeting key biochemical characteristics, e.g. lactose fermentation and indole production. Selected additional biochemical tests may be required for complete identification of the organism (Table 1.2).

In more recent years other characteristics of the organism have been exploited in media and method developments. In particular, the characteristic of most *E. coli* to produce β-glucuronidase is used in a range of chromogenic (colour producing) and fluorogenic (fluorescence under ultra-violet light) media that are now available (Tables 6.2 and 6.3) which, together with a simple test for indole production, allow the confirmed detection of *E. coli*.

Some of the classical biotypic characteristics by which *E. coli* is usually identified (Table 1.2), particularly in food industry isolation and identification procedures, have been found to be unreliable for the detection of some groups of pathogenic *E. coli*. For example, many strains of VTEC O157 do not ferment D-sorbitol in 24 h, do not ferment rhamnose, do decarboxylate lysine and ornithine, do ferment raffinose and dulcitol and do not produce β-glucuronidase. In addition, some strains have an upper limit for growth of 41°C in selective media (Chapman, 1995) so the industry approach to detecting *E. coli* using an incubation temperature of 44°C as part of the selective isolation procedure (Figure 6.1) is ineffective for the isolation of some important strains of VTEC. Most strains of EIEC are non-motile, lysine decarboxylase negative and may be lactose-negative (Hale *et al.*, 1997). This has led to the development of different methods for isolating specific pathogenic types of *E. coli* from foods and characterizing the isolates; particularly for *E. coli* O157:H7.

A great deal of work has been carried out in recent years, particularly in public health and clinical laboratories, to ascertain the most productive and selective methods for isolating and identifying *E. coli* O157:H7. Food industry microbiologists have essentially followed the lead of clinical colleagues and the current method of choice for the detection of *E. coli* O157:H7 is one based on the conventional use of selective enrichment and plating media but employing immunomagnetic beads for selective capture of the target organism (Figure 6.2) (Anon., 1997h).

It is clear that all aspects of media and methods for the detection and identification of *E. coli* and pathogenic *E. coli* will continue to be developed and improved with respect to reliability, sensitivity, selectivity and speed of obtaining results.

One particular aspect of working with *E. coli* O157:H7 should be noted. Potentially pathogenic *E. coli* are currently classified in the UK by the Advisory Committee on Dangerous Pathogens in Hazard Group 2 (Advisory Committee on Dangerous Pathogens, 1995). Because of the particularly severe hazard presented by *E. coli* O157:H7 and other VTECs, the low infective doses required to cause illness and the reports of laboratory-acquired infections by these organisms, the hazard categorization of VTECs is under review. It is expected that these will be recategorized to Hazard Group 3 and relevant health and safety documentation will be published by UK

Table 6.2 Examples of alternative methods available for the detection and/or identification of *E. coli*, *E. coli* O157 and *E. coli* O157:H7

Test type	*E. coli* or VTEC O157	Name of test	Supplier
Chromogenic and/or fluorogenic media	*E. coli*	Petrifilm *E. coli* Count Plate	3M Health Care
	E. coli	Colilert	Idexx Laboratories Inc.
	E. coli	Tryptone Bile Glucuronide Agar	LAB M, International Diagnostics Group plc
	E. coli	Fluorocult ECD Agar	Merck Ltd
	E. coli	TBX Medium	Oxoid Ltd
	E. coli	Chromogenic *E. coli*/coliform Medium	Oxoid Ltd
	E. coli O157:H7	Fluorocult *E. coli* O157:H7 Agar	Merck Ltd
	E. coli O157:H7	Rainbow Agar O157	Biolog Inc.
Miniaturized biochemical test kits	*E. coli*	API 20E	bioMérieux UK Ltd
		VITEK	bioMérieux UK Ltd
		Micro-ID	Organon Teknika Ltd
Electrical	*E. coli*	Bactometer	bioMérieux UK Ltd
	E. coli	Malthus	Malthus Instruments
	E. coli	RABIT	Don Whitley Scientific Ltd
Enzyme-linked immunosorbent assay (ELISA)	*E. coli* O157	Vidas *E. coli* O157	bioMérieux UK Ltd
	E. coli O157	EHEC-Tek	Organon Teknika Ltd
	E. coli O157	TECRA *E. coli* O157 Visual Assay	TECRA Diagnostics
	E. coli O157	*E. coli* O157 Rapitest	Microgen Bioproducts Ltd
	E. coli O157	Petrifilm-HEC	3M Healthcare
	E. coli O157	EZ Coli Rapid Detection System	Difco Laboratories Ltd
	E. coli O157:H7	Assurance EIA	M-Tech Diagnostics Ltd

Table 6.2 Continued

Test type	*E. coli* or VTEC O157	Name of test	Supplier
Immuno-chromatography	*E. coli* O157:H7	Reveal for *E. coli* O157:H7	RossLab plc
	E. coli O157:H7	Reveal 8 for *E. coli* O157:H7	RossLab plc
	E. coli O157:H7	Visual Immunoprecipitate Assay	M-Tech Diagnostics Ltd Bio Control Systems Inc.
	E. coli O157	C QUIC Plus *E. coli* O157 test	Sun International Trading
Immuno-capture	*E. coli* O157	Dynabead O157	Dynal Ltd
	E. coli O157	*E. coli* O157 Immunocapture Confirmation System	TECRA Diagnostics
Latex agglutination	*E. coli* O157	*E. coli* O157 Latex	Oxoid Ltd
	Vero cytotoxigenic *E. coli*	VTEC-RPLA	Oxoid Ltd
	E. coli O157	MicroScreen *E. coli* O157	Microgen Bioproducts Ltd
	E. coli O157:H7	Pro-Lab *E. coli* O157 Latex Test Reagent Kit	Pro-Lab Diagnostics
Nucleic acid hybridization probe	*E. coli*	Gene Trak	Gene Trak Systems
Polymerase chain reaction	*E. coli* O157:H7	BAX system	Qualicon, Dupont

Table 6.3 Practical aspects associated with the use of some conventional and alternative methods for the detection and identification of *E. coli*, *E. coli* O157 and *E. coli* O157:H7

Test	Presumptive/ confirmed result	Further confirmation/ identification	Approximate time to carry out the specific test	Approximate total time to negative result from initial sampling*	Approximate total time to confirmed positive result from initial sampling*
Chromogenic and/or fluorogenic media					
Petrifilm *E. coli* Count Plate	Confirmed *E. coli*	None	24–48 h	24–48 h	24–48 h
Colilert	Confirmed *E. coli*	None	18–24 h	18–24 h	18–24 h
Tryptone Bile Glucuronide Agar	Presumptive *E. coli*	Yes	22 h	22 h	22 h
Fluorocult ECD Agar	Presumptive *E. coli*	Yes	18–24 h	18–24 h	72 h
TBX Medium	Confirmed *E. coli*	None	24 h	24 h	24 h
Fluorocult *E. coli* O157:H7 Agar	Presumptive *E. coli* O157:H7	Yes	48 h	48 h	2–3 days
Rainbow Agar O157	Presumptive *E. coli* O157:H7	Yes	24–48 h	24–48 h	2–3 days
Miniaturized biochemical test kits	Confirmed *E. coli*	None	24 h	72–96 h	Up to 5 days
Electrical					
Bactometer	Presumptive *E. coli*	Yes	< 48 h	< 48 h	4–5 days
Malthus	Presumptive *E. coli*	Yes	< 48 h	< 48 h	4–5 days
RABIT	Presumptive *E. coli*	Yes	< 48 h	< 48 h	4–5 days

Table 6.3 Continued

Test	Presumptive/ confirmed result	Further confirmation/ identification	Approximate time to carry out the specific test	Approximate total time to negative result from initial sampling*	Approximate total time to confirmed positive result from initial sampling*
Enzyme-linked immunosorbent assay (ELISA)					
Vidas *E. coli* O157	Presumptive *E. coli* O157	Yes	45 min	25 h	3–4 days
EHEC-Tek	Presumptive *E. coli* O157	Yes	3 h	30 h	3–4 days
TECRA *E. coli* O157 Visual Assay	Presumptive *E. coli* O157	Yes	2 h	20 h	3–4 days
E. coli O157 Rapitest	Presumptive *E. coli* O157	Yes	< 4 h	28 h	3–4 days
Petrifilm-HEC	Presumptive *E. coli* O157	Yes	18 h	28 h	3–4 days
EZ Coli Rapid Detection System	Presumptive *E. coli* O157	Yes	< 20 min	24 h	3–4 days
Immunochromatography					
Reveal for *E. coli* O157:H7	Confirmed *E. coli* O157:H7	None	15 min	24 h	24 h
Reveal 8 for *E. coli* O157:H7	Confirmed *E. coli* O157:H7	None	15 min	8 h	8 h
VIP assay	Confirmed *E. coli* O157:H7	None	10 min	24 h	24 h

Table 6.3 Continued

Test	Presumptive/ confirmed result	Further confirmation/ identification	Approximate time to carry out the specific test	Approximate total time to negative result from initial sampling*	Approximate total time to confirmed positive result from initial sampling*
C QUIC plus *E. coli* O157 test	Presumptive *E. coli* O157	Yes	15 min	28 h	4 days
Latex agglutination					
E. coli O157 latex	Confirmed *E. coli* O157	Only if serotyping required	2 min	24 h	24 h
VTEC-RPLA	Confirmed VT1 and VT2 of VTECs	Only if serotyping required	24 h	48–72 h	48–72 h
MicroScreen *E. coli* O157	Confirmed *E. coli* O157	Only if serotyping required	15 min	24 h	24 h
Pro-Lab *E. coli* O157 latex test reagent kit	Presumptive *E. coli* O157:H7	Yes	2 min	24 h	72 h
Polymerase chain reaction	BAX system confirmed *E. coli* O157:H7	None	3.5 h	28 h	28 h

* Includes time which may be required for any associated culture work e.g. prior to conducting the test, purification of suspect positive colonies/cultures prior to conducting relevant confirmatory tests and the confirmatory tests.

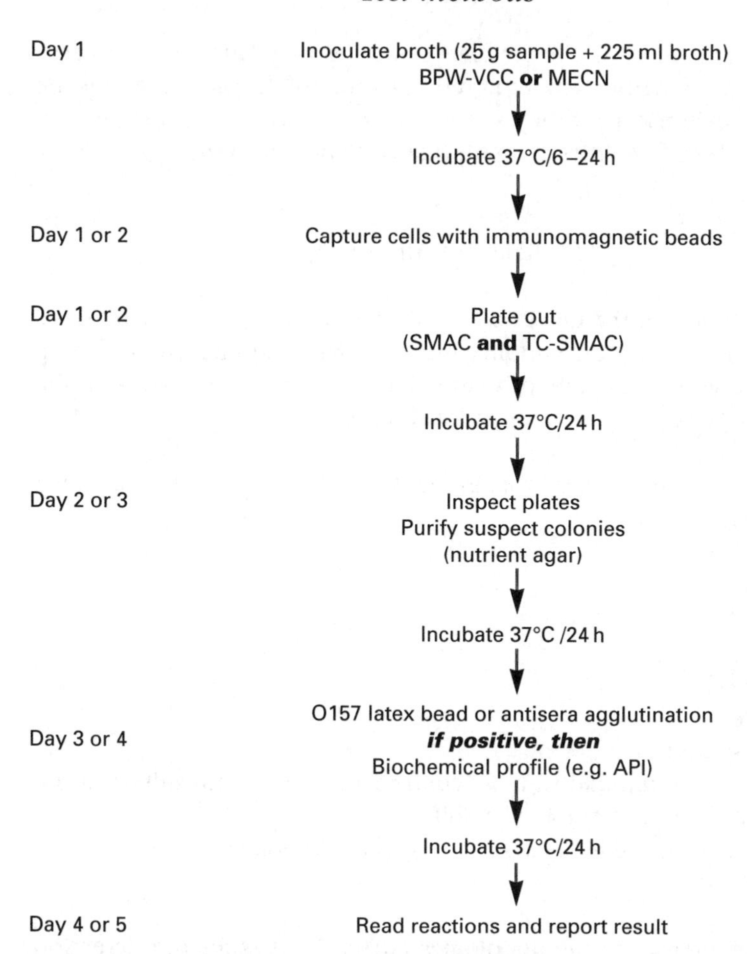

Figure 6.2 Current common procedure for the detection and identification of *E. coli* O157 in the UK. BPW-VCC = buffered peptone water + Vancomycin (8 mg/l final concentration), Cefixime (0.05 mg/l final concentration) and Cefsulodin (10 mg/l final concentration); MECN = modified EC broth + Novobiocin (20 mg/l final concentration); SMAC = sorbitol MacConkey agar; TC-SMAC = sorbitol MacConkey agar + potassum tellurite (2.5 mg/l final concentration) and Cefixime (0.05 mg/l final concentration).

government agencies in Spring 1998. Such a recategorization may preclude many food microbiology laboratories from carrying out tests for these organisms as they do not have the containment facilities necessary for handling Hazard Group 3 organisms (Advisory Committee on Dangerous Pathogens, 1995). Indeed, the UK Health and Safety Executive (Health and Safety Executive, 1996) issued some interim advice concerning laboratories working with VTEC strongly advising risk assessment under legislation applying to the Control of Substances Hazardous to Health. With this in mind, it may be advisable for many food industry laboratories to seriously consider whether any required tests for this organism should be carried out 'in-house' or if they should be contracted out to a laboratory with the necessary facilities and staff trained to deal with these organisms.

ALTERNATIVE METHODS

As with methods for the detection of other foodborne bacteria, there is a need to supplement conventional methods with simpler, labour-saving methods preferably capable of delivering results more quickly and reliably than is possible using conventional methods.

When selecting any method for use all of the following attributes need consideration:

- sensitivity
- specificity
- simplicity
- robustness
- reliability
- 'hands-on' time
- the need for additional tests to confirm presumptive results and the time taken to obtain the final result
- requirement for trained staff and/or special equipment
- cost per test.

A variety of alternative methodologies currently exist for the detection and/or identification of *E. coli* and *E. coli* O157:H7 (Table 6.2) and the list is growing rapidly as new technologies are exploited for application based on the needs of microbiologists. Regular updates on commercial alternative microbiological test methods are available together with useful references and the validation status of the method (Betts, 1997).

Many alternative tests depend on the presence of a high number of target cells for detection and the procedures specified for use with a manufacturer's kit or test are designed to assure this level is reached. To minimize false negative or false positive results it is important to read and understand the manufacturer's technical information supplied, and carefully follow the instructions for carrying out the test.

The following are some of the alternative test methods more commonly used in routine testing laboratories. It is important to recognize that the enormous amount of commercial effort still ongoing to develop new and alternative methods for detecting *E. coli*, VTEC and *E. coli* O157/O157:H7 will add yet further tests and methods to the list. In selecting methods for use, a clear understanding of the test's capability is necessary. Of particular importance is the actual target organism, which from the stated aims of currently available tests could be *E. coli*, *E. coli* O157, *E. coli* O157:H7 or Vero cytotoxigenic *E. coli*. Table 6.3 summarizes some of the practical information concerning the use of some of the technologies applied for the specific isolation and/or identification of *E. coli* and specified types of *E. coli*.

When conducting tests for confirming the identity of *E. coli* and/or *E. coli* O157:H7 it is good practice to use purified cultures. Where applicable, the time necessary for the production of pure cultures has been included in the total approximate times indicated in Table 6.3 to produce a result from the examination of presumptive positive broths or colonies.

Biochemical identification kits

These are some of the simplest forms of labour-saving test systems for identifying *E. coli*. They consist of a range of biochemical reactions produced in preformed chambers supplied in a disposable unit. Following inoculation and incubation of the test chambers, reactions are assessed usually by a specified colour change in the medium. Following assessment, a profile of the organism is obtained, which is used to determine the identity of the *E. coli*. For the specific identification of *E. coli* O157:H7, it is necessary to supplement the test results with separate tests using selective media and serology.

Enzyme-linked immunosorbent assays (ELISA)

ELISA tests are based on the high specificity of an antibody to its target antigen; in this case, *E. coli* O157. Some tests claim to employ antibodies to *E. coli* O157:H7 and it is therefore important to understand a test's capability with respect to the level of identification achieved. The anti-O157 antibodies are bound to a solid substrate, e.g. the internal surface of the wells in a microtitre plate, and these are used to capture O157 antigens present in the treated enrichment broth placed in the well. Following a sequence of manipulations involving washings, addition of further reagents and incubations, a coloured end product is obtained in those wells containing *E. coli* O157. Sample enrichment based on the use of conventional selective enrichment broths is necessary to ensure sufficient target cells are available (10^5–10^6 per ml) for detection of a contaminated sample.

Positive results obtained using ELISAs are usually confirmed by streaking the original enrichment broth onto selective agar and then following the conventional approach to confirm the presence and specific type of *E. coli*.

ELISAs can offer rapid screening of samples with a negative result available in 24 h and a potential early indication of a positive result.

Immunomagnetic separation

This technique is used to improve the sensitivity and specificity of conventional selective methods for detecting the presence of *E. coli* O157 in samples. It employs small (<100 μm) magnetic beads coated with antibodies to *E. coli* O157. The beads are used to specifically capture cells of the organism from enrichment broths. The magnetic property of the beads is used to separate the cells from the broth and these are plated out onto selective agar and the conventional approach then followed to identify the specific *E. coli* type. This application of immunomagnetic beads currently forms part of the most commonly used method for the detection of *E. coli* O157 in foods.

Latex agglutination tests

Antibody-coated latex beads are used in agglutination tests to provide a rapid indication of the presence of colonies of *E. coli* O157 or VTECs on selective agar culture plates. Such tests offer specificity and time saving, which can be important to a food microbiologist examining products against specifications.

OTHER METHODS

A variety of other alternative methods are currently either not yet widely used in food industry laboratories or only used in specific sectors of industry is available.

Immunochromatography

In these tests labelled antibodies to *E. coli* O157 are located in a 'reaction' zone on a chromatographic pad unit. Generally, cultured enrichment broth is added to a sample point at the bottom of the unit and migrates through the membrane. The immobilized labelled antibody binds to the target organism to form a line of detection in the test 'window' indicating a positive result. Like latex agglutination tests, immunochromatographic tests may offer specificity and time saving, which can be important to a food microbiologist examining products against specifications.

Electrical techniques

Electrical techniques for the detection of microorganisms are based on the ability of instruments to monitor and detect small changes in the electrical properties of a medium in which a microorganism is growing.

Specific broth media have been developed that favour the growth of *E. coli*. As the organism grows, particular substrates in the broth are utilized, yielding products of metabolism giving detectable changes in conductance. The changes occurring are recorded as a conductance curve. The curve generated is monitored by computer; if it meets certain pre-set criteria the sample is registered as positive and the broth is considered to be presumptive positive for *E. coli*. Confirmation of any presumptive positive broths is carried out by streaking from the broth onto selective agar and then following the conventional approach to identify the organism present.

'Fingerprinting' methods

Once *E. coli* has been isolated and identified using conventional methods, this level of identity together with the associated sample information is sufficient for the food microbiologist to interpret results against the original requirements for the test, e.g. a product-buying specification. However, because of the increasing concern over the implications of VTECs in foods, it may be necessary to further discriminate the identity of the strain isolated, particularly if the origin of the strain needs to be traced in connection with cases of suspected illness.

The ability to subtype or 'fingerprint' *E. coli* serogroups is important in the surveillance of pathogenic *E. coli* as well as in the investigation of outbreaks of the illness and traceability in food processing and environmental/ecological situations. Techniques are available that can 'fingerprint' pathogenic strains of *E. coli*, allowing confident traceability of strains in the factory environment. Methods used for subclassifying *E. coli* include biotyping, serotyping, phage typing, colicin typing, typing according to outer membrane protein pattern, antibiotic resistance patterns (resistotyping), direct haemagglutination, profile of intracellular enzymes, plasmid profile and production of enterotoxins. In particular, tests for motility, VT type, phage type and pulsed-field gel electrophoresis (PFGE) pattern are already used in practical schemes for differentiating strains of VTEC O157 (Advisory Committee on the Microbiological Safety of Food, 1995; Pennington, 1997). Should this level of identity be required, then the services of a national public health sector laboratory expert in these techniques and the interpretation of the results should be obtained.

It is always important to ensure that the method used will reliably detect the *E. coli* of concern (if it is present) in the food being examined. It may be necessary to initially confirm by experiment (validate) that a particular method will be suitable for the purpose intended. Subsequently, in routine use, quality assurance systems (internal and external) should be employed to verify ongoing test efficacy.

A variety of sources exist that may be used for guidance in appropriate method selection. National and international standard methods published, for example, by the British Standards Institution, the International Standards Organization and the International Dairy Federation are available. In addition, methods have been reviewed, practised and validated by reputable bodies such as the Public Health Laboratory Service (UK), the Association of Official Analytical Chemists (USA) and Campden and Chorleywood Food Research Association (UK).

7

THE FUTURE

Despite the fact that a vast amount of work has been carried out on all aspects of *E. coli* during the 100 years or more since it was first described, the organism continues to provide new challenges because of the wide diversity of types within the species, ranging from harmless commensals to dangerous human pathogens.

A great deal more work has yet to be done to build our knowledge and understanding of the routes of transmission to humans of the many different pathogenic serotypes of the organism. From such a knowledge base, the food industry can develop and maintain control procedures to prevent these serotypes from being a risk to human health from food sources.

Although *E. coli* O157:H7 is currently the most predominant human VTEC infection (and is also foodborne), there are other foodborne pathogenic VTEC serotypes. Information concerning their particular food sources, growth and survival characteristics, detection methods and effective, practical control measures that can be applied in the food industry is required and must form a part of both short- and mid-term future work programmes.

Over the past decade in particular, commercial considerations have driven food product development to become increasingly innovative. The ready availability of food raw materials from anywhere in the world leading to the use of more 'exotic' ingredients and the closure of the 'season gap', which has made normally seasonal foods available all year round, have given free rein to innovative food technology.

Developments in such raw materials availability and novel applications, combined with developments in processing technologies, packaging technology and storage and distribution systems, are all providing continuing challenges to food microbiologists and technologists to devise and maintain controls of food safety.

The market for chilled food products, including savoury ready meals (recipe dishes), dairy and dessert products, sandwiches and other snack meals, has grown extensively over the last decade. Increasingly novel and complex combinations of raw materials are being manufactured in large-scale commercial systems using the wide variety of meats, poultry, game, fish and shellfish, milk and milk products, eggs, vegetables (root, leafy and salad), grains, nuts, herbs and spices available.

Alternative food processing technologies are in development and new applications of existing technologies are being explored for use in food production processes. Ohmic heating processes, irradiation, ultra-filtration, high pressure and high intensity light are examples of such technologies. The effects of these on the survival of microorganisms in respect of food safety and food spoilage have, in many instances, yet to be determined. Clearly, the list of already known bacterial pathogens, including the pathogenic *E. coli*, must be included in any consideration of the implications of new combinations of approaches to food product development and the application of alternative technologies in food processing.

From the outbreaks and survey information already available, some of which are described in this book, it is already clear that the primary production environments of many food industry raw materials, particularly meat, vegetables and salad vegetables, contribute greatly to a widespread low incidence of bacterial pathogens, including *E. coli* O157, in these raw materials.

It is inevitable that, as more work is carried out, the list of sources of the organism will extend. It is therefore essential that a detailed and competent hazard analysis is carried out at an early stage in all new food product and process developments to ensure that relevant critical controls and monitoring systems can be put in place. This will help to minimize potential public health problems that could arise from the presence and outgrowth of all types of pathogenic *E. coli*, especially the Vero cytotoxigenic types.

GLOSSARY OF TERMS

Biotyping The conventional method for distinguishing between bacterial types using their metabolic and/or physiological properties (biotype).

Commensal Animals or plants that live as tenants of others and share their food.

***D* value** The time required (usually expressed in minutes) at a given temperature to reduce the number of viable cells or spores of a given microorganism to 10% of the initial population.

Food poisoning Any disease of an infectious or toxic nature caused by the consumption of food or water.

Genotyping Methods used to differentiate bacteria based on the composition of their nucleic acids.

HACCP (Hazard Analysis Critical Control Point) The systematic approach to the identification and assessment of the hazards and risks in a food production process and the determination of means for their control for assuring food safety.

Humectant A substance that absorbs moisture.

Indicator organisms Those organisms whose presence suggests inadequate processing for safety.

Nephropathy Kidney disorder.

Pasteurization A form of heat treatment that kills vegetative pathogens and spoilage microorganisms in milk and other foods: for example, for milk a common pasteurization process is 71.7°C for 15 s.

Pathogen Any microorganism that by direct interaction with (infection of) another organism causes disease in that organism.

Phage typing A method used to distinguish between bacteria within the same species on the basis of their susceptibility to a range of bacterial viruses (bacteriophage).

Phenotype The observable characteristics of an organism, which include biotype, serotype, phage type and bacteriocin type.

Polymerase chain reaction (PCR) A technique used to amplify the number of copies of a preselected region of DNA to a sufficient level for testing.

Pulsed-field gel electrophoresis (PFGE) A technique that allows chromosomal restriction fragment patterns to be produced.

Ribotyping A method for characterizing bacterial isolates according to their ribosomal RNA pattern (ribotype) and identifying the isolate by comparing the pattern obtained to a database of patterns.

Serotyping A method of distinguishing bacteria on the basis of their antigenic properties (reaction to known antisera). The O antigen defines the *serogroup* of a strain and the H antigen defines the *serotype* of the strain, therefore a number of serotypes may constitute a serogroup.

Strain An isolate or group of isolates that can be distinguished from other isolates of the same genus and species by phenotypic and/or genotypic characteristics.

Thrombocytopaenia Low numbers of platelets circulating in the bloodstream.

Thrombotic thrombocytopaenic purpura (TTP) Condition which results from the aggregation of platelets in various organs; characterised by fever, anaemia and kidney failure.

Vero cells African green monkey kidney cells used in tissue culture.

Vero cytotoxin Toxin harmful to cultured Vero cells.

Water activity (a_w) A measure of the availability of water for the growth and metabolism of microorganisms. It is expressed as a ratio of the water vapour pressure of a food or solution to that of pure water at the same temperature.

***z* value** The change in temperature (°C) required to change the *D* value 10-fold.

REFERENCES

Abbar, F.M. (1988) Incidence of fecal coliforms and serovars of enteropathogenic *Escherichia coli* in naturally contaminated cheese. *Journal of Food Protection*, **51**(5), 384–385.

Abdul-Raouf, U.M., Beuchat, L.R. and Ammar, M.S. (1993) Survival and growth of *Escherichia coli* O157:H7 on salad vegetables. *Applied and Environmental Microbiology*, **59**(7), 1999–2006.

Adams, M.R., Hartley, A.D. and Cox, L.J. (1989) Factors affecting the efficacy of washing procedures used in the production of prepared salads. *Food Microbiology*, **6**, 69–77.

Advisory Committee on Dangerous Pathogens (1995) *Categorisation of Biological Agents According to Hazard and Categories of Containment,* 4th edn. HSE Books, Sudbury, Suffolk.

Advisory Committee on the Microbiological Safety of Food (1995) *Report on Vero Cytotoxin-Producing* Escherichia coli, HMSO, London.

Ahmed, S. (1997) An outbreak of *E. coli* O157 in Central Scotland. *Scottish Centre for Infection and Environmental Health Weekly Report*, Number 1, No. 97/13, 8. Scottish Centre for Infection and Environmental Health, Glasgow.

Ahmed, N.M., Conner, D.E. and Huffman, D.L. (1995) Heat-resistance of *Escherichia coli* O157:H7 in meat and poultry as affected by product composition. *Journal of Food Science*, **60**(3), 606–610.

Alexander, E.R., Boase, J., Davis, M., *et al.* (1995) *Escherichia coli* O157:H7 outbreak linked to commercially distributed dry-cured salami – Washington and California, 1994. *Morbidity and Mortality Weekly Report*, **44**(9), 157–160.

Anon. (1992a) *Safer Cooked Meat Production Guidelines. A 10 point plan.* Department of Health, London.

Anon. (1992b) Council Directive 92/46/EEC of 16th June 1992 laying down the health rules for the production and placing on the market of raw milk, heat-treated milk and milk-based products. *Official Journal of the European Communities*, 14.9.92, No. L268, 1–34.

Anon. (1993a) Cooking ground beef. Memorandum from the Department of Health and Human Services, January 27, USA.

Anon. (1993b) Council Directive 93/43/EEC of 14th June 1993 on the hygiene of foodstuffs. *Official Journal of the European Communities*, 19.7.93, No. L175, 1–11.

Anon. (1993c) Food poisoning – *E. coli* O157. *MAFF Food Safety Directorate Information Bulletin*, No. 35, March, 1.

Anon. (1994a) Two clusters of haemolytic uraemic syndrome in France. *Communicable Disease Report*, **4**(7), 1, Public Health Laboratory Service, London.

Anon. (1994b) *Nationwide Beef Microbiological Baseline Data Collection*

Program: Steers and Heifers, October 1992-September 1993. United States Department of Agriculture, Food Safety Inspection Service, Washington, DC.
Anon. (1994c) *Microbiological Examination of Food and Animal Feeding Stuffs. Part 8: Enumeration of Presumptive* Escherichia coli. *Most Probable Number Technique*. BS 5763 British Standards Institution, London (ISO 7251:1993).
Anon. (1995a) The prevention of human transmission of gastrointestinal infections, infestations, and bacterial intoxications, *Escherichia coli*-Vero cytotoxin producing (VTEC). *Communicable Disease Report*, **5**, Review No. 11, R164, Public Health Laboratory Service, London.
Anon. (1995b) Interim guidelines for the control of infections with Vero cytotoxin-producing *Escherichia coli* (VTEC). *Communicable Disease Report*, **5**, Review No. 6, R77-R81, Public Health Laboratory Service, London.
Anon. (1995c) Garibaldi food poisoning outbreak coroner's report. *Supplement to Environmental Health Service Bulletin*, No. 15, November 1995, 1-8.
Anon. (1995d) *The Food Safety (General Food Hygiene) Regulations 1995*. Statutory Instrument No. 1763, HMSO, London.
Anon. (1995e) *Report on the National Study of Ready-to-eat Meats and Meat Products, Part 1*. Ministry of Agriculture, Fisheries and Food, London.
Anon. (1996a) *Backgrounder: Be a part of reducing the risk of illness from* E. coli *O157:H7 - Government, health, business and consumers working together*. US Department of Agriculture, USA.
Anon. (1996b) *Dry Fermented Sausage and* E. coli *O157:H7*, Research Report No. 11. Blue Ribbon Task Force, National Cattlemen's Association, Chicago, IL.
Anon. (1996c) IFST Position Statement on Verocytotoxin-producing *E. coli* (VTEC). *Keynote, Newsletter of the Institute of Food Science and Technology (UK)*, November, 4-5.
Anon. (1996d) *Report on the National Study of Ready-to-eat Meats and Meat Products, Part 3*. Ministry of Agriculture, Fisheries and Food, London.
Anon. (1996e) *Report on the National Study of Ready-to-eat Meats and Meat Products, Part 2*. Ministry of Agriculture, Fisheries and Food, London.
Anon. (1996f) Microbiological guidelines for some ready-to-eat foods sampled at the point of sale: an expert opinion from the Public Health Laboratory Service. *PHLS Microbiology Digest*, **13**(1), 41-43.
Anon. (1996g) *Proposed Draft: Revised Principles for the Establishment and Application of Microbiological Criteria for Foods*, ALINORM 97/13A, Annex to Appendix III. Codex Committee on Food Hygiene, Codex Alimentarius Commission, Rome.
Anon. (1997a) *PHLS Vero Cytotoxin-Producing* Escherichia coli *O157 Fact Sheet*. Public Health Laboratory Service, London.
Anon. (1997b) *Report on the National Study on Ready-to-eat Meats and Meat Products, Part 4*. Ministry of Agriculture, Fisheries and Food, London.
Anon. (1997c) Vero cytotoxin-producing *Escherichia coli* O157. *Communicable Disease Report*, **7**, 46, 412, Public Health Laboratory Service, London.
Anon. (1997d) *Guidelines for Good Hygienic Practice in the Manufacture of Chilled Foods*, 3rd edn. Chilled Food Association, London.
Anon, (1997e) *Draft Hazard Analysis and Critical Control Point (HACCP) System and Guidelines for its Application*. ALINORM 97/13A, Appendix II. Codex Committee on Food Hygiene. Codex Alimentarius Commission, Rome.
Anon. (1997f) Development and use of microbiological criteria for foods. *Food Science and Technology Today*, **11**(3), 137-177.
Anon. (1997g) *Milk and Milk Products - Enumeration of Presumptive*

Escherichia coli. *Part 1: Most Probable Number Technique* and *Part 3: Colony-count Technique at 44°C using Membranes.* BS ISO 11866, British Standards Institution, London.

Anon. (1997h) *Manual of Microbiological Methods for the Food and Drinks Industry, Technical Manual No. 43*, 2nd edn, Methods 5.1, 5.2 and 19.1. Campden and Chorleywood Food Research Association, Chipping Campden.

Anon. (1998a) Common gastrointestinal tract infections, England and Wales: Laboratory reports, weeks 01-52, 1997. *Communicable Disease Report*, **8**(2), 14.

Anon. (1998b) Common gastrointestinal tract infections, England and Wales: Laboratory reports, weeks 01-06, 1998. *Communicable Disease Report*, **8**(7), 59.

Arocha, M.M., Mcvey, M., Loder, S.D. *et al.* (1992) Behaviour of hemorrhagic *Escherichia coli* O157:H7 during the manufacture of cottage cheese. *Journal of Food Protection*, **55**(5), 379-381.

Ault, A. and Morris, K. (1997) USA acts to contain potential *E. coli* disaster. *The Lancet*, 350 (August 23), 567.

Australia New Zealand Food Authority (1996a) *Requirements for making uncooked fermented comminuted meat products from 27 June 1996.* Australia New Zealand Food Authority, Canberra.

Australia New Zealand Food Authority (1996b) *Your Food, Your Safety.* Australia New Zealand Food Authority, Canberra.

Bachmann, H.P. and Spahr, U. (1995) The fate of potentially pathogenic bacteria in Swiss hard and semihard cheeses made from raw milk. *Journal of Dairy Science*, **78**, 476-483.

Bell, B.P., Goldoft, M., Griffin, P.M. *et al.* (1994) A multistate outbreak of *Escherichia coli* O157:H7 - associated bloody diarrhea and hemolytic uremic syndrome from hamburgers. The Washington Experience. *Journal of the American Medical Association*, **272**(17), 1349-1353.

Belongia, E.A., MacDonald, K.L., Parham, G.L. *et al.* (1991) An outbreak of *Escherichia coli* O157:H7 colitis associated with consumption of precooked meat patties. *The Journal of Infectious Diseases*, **164**, 338-343.

Besser, R.E., Lett, S.M., Weber, J.T. *et al.* (1993) An outbreak of diarrhea and hemolytic uremic syndrome from *Escherichia coli* O157:H7 in fresh-pressed apple cider. *Journal of the American Medical Association*, **269**(17), 2217-2220.

Bettelheim, K.A. (1997) *Escherichia coli* in the normal flora of humans and animals, in Escherichia coli: *Mechanisms of Virulence* (ed. M. Sussman). Cambridge University Press, Cambridge, pp. 85-109.

Betts, G.D., Lyndon, G. and Brooks J. (1993) Heat resistance of emerging foodborne pathogens: *Aeromonas hydrophila*, *Escherichia coli* O157:H7, *Plesiomonas shigelloides* and *Yersinia enterocolitica*. Technical Memorandum No. 672, Campden and Chorleywood Food Research Association, Chipping Campden.

Betts, R.P. (1997) *The Catalogue of Rapid Microbiological Methods, Review No. 1*, 3rd edn. Campden and Chorleywood Food Research Association, Chipping Campden.

Beuchat, L.R. (1992) Surface disinfection of raw produce. *Dairy, Food and Environmental Sanitation*, **12**(1), 6-9.

Beuchat, L.R. (1996) Pathogenic microorganisms associated with fresh produce. *Journal of Food Protection*, **59**(2), 204-216.

Beutin, L., Geier, D., Steinrück, H. *et al.* (1993) Prevalence and some properties of verotoxin (Shiga-like-toxin) producing *Escherichia coli* in seven different species of healthy domestic animals. *Journal of Clinical Microbiology*, **31**, 2483–2488.

Beutin, L., Gleier, K., Kontny, I. *et al.* (1997) Origin and characteristics of entero-invasive strains of *Escherichia coli* (EIEC) isolated in Germany. *Epidemiology and Infection*, **118**, 199–205.

Bolton, F.J., Crozier, L. and Williamson, J.K. (1996) Isolation of *Escherichia coli* O157 from raw meat products. *Letters in Applied Microbiology*, **23**, 317–321.

Bramley, A.J. and McKinnon, C.H. (1990) The microbiology of raw milk, in *Dairy Microbiology: The Microbiology of Milk*, 2nd edn. (ed. R.K. Robinson). Elsevier Applied Science, London.

Brenner, D.J. (1984) Family 1. Enterobacteriaceae RAHN 1937, Nom. Fam. Cons. Opin. 15, Jud. Comm. 1958, 73; Ewing, Farmer, and Brenner 1980, 674; Judicial Commission 1981, 104, in *Bergey's Manual of Systematic Bacteriology*, vol. 1. Williams & Wilkins, Baltimore, p. 408.

Cameron, A.S., Beers, M.Y., Walker, C.C. *et al.* (1995a) Community outbreak of hemolytic uremic syndrome attributable to *Escherichia coli* O111:NM – South Australia, 1995. *Morbidity and Mortality Weekly Report*, July 28, 550–551, 557–558.

Cameron, S., Walker, C., Beers, M. *et al.* (1995b) Enterohaemorrhagic *Escherichia coli* outbreak in South Australia associated with the consumption of Mettwurst. *Communicable Disease Intelligence,* **19**(3), 70–71.

Chapman, P.A. (1995) Verocytotoxin-producing *Escherichia coli*: an overview with emphasis on the epidemiology and prospects for control of *E. coli* O157. *Food Control*, **6**(4), 187–193.

Chapman, P.A., Siddons, C.A., Wright, D.J. *et al.* (1993) Cattle as a possible source of Verocytotoxin-producing *Escherichia coli* O157 infections in man. *Epidemiology and Infection*, **111**, 439–447.

Chapman, P.A., Siddons, C.A. and Harkin, M.A. (1996) Sheep as a potential source of Verocytotoxin-producing *Escherichia coli* O157. *The Veterinary Record*, January 6, 23–24.

Coia, J.E. and Hanson, M.F. (1997) A survey of the prevalence of *E. coli* O157 (VTEC) in raw meats, raw cows' milk and raw milk cheeses in south-east Scotland. *Scottish Centre for Infection and Environmental Health Weekly Report*, **13**(1), 17–18.

Como-Sabetti, K., Reagan, S., Allaire, S. *et al.* (1997) Outbreaks of *Escherichia coli* O157:H7 infection associated with eating alfalfa sprouts – Michigan and Virginia, June–July 1997. *Morbidity and Mortality Weekly Report*, **46**(32), 741–744.

Corry, J.E.L., James, C., James, S.J. *et al.* (1995) *Salmonella*, *Campylobacter* and *Escherichia coli* O157:H7 decontamination techniques for the future. *International Journal of Food Microbiology*, **28**, 187–196.

Curnow, J. (1994) *E. coli* O157 phage type 28 infections in Grampian. *Communicable Diseases and Environmental Health of Scotland*, **28**(94/46), 1.

Cutter, C.N., Dorsa, W.J. and Siragusa, G.R. (1996) Application of Carnatrol™ and Timsen™ to decontaminate beef. *Journal of Food Protection*, **59**(12), 1339–1342.

Dargatz, D.A., Wells, S.J., Thomas, L.A. *et al.* (1997) Factors associated with the presence of *Escherichia coli* O157 in feces of feedlot cattle. *Journal of Food Protection*, **60**(5), 466–470.

Davis, H.J.P., Taylor, J.N., Perdue, G.N. *et al.* (1988) A shigellosis outbreak traced to commercially distributed lettuce. *American Journal of Epidemiology*, **128**, 1312–1321.

Davis, M., Osaki, C., Gordon, D. *et al.* (1993) Update: Multistate outbreak of *Escherichia coli* O157:H7 infection from hamburgers – Western United States 1992–1993. *Morbidity and Mortality Weekly Report*, **42**, 258–263.

De Louvois, J. and Rampling, A. (1998) One fifth of samples of unpasteurised milk are contaminated with bacteria. *British Medical Journal*, **316**, 21 Feb 1998, 625.

Del Rosario, B.A. and Beuchat, L.R. (1995) Survival and growth of enterohemorrhagic *Escherichia coli* O157:H7 in cantaloupe and watermelon. *Journal of Food Protection*, **58**(1), 105–107.

Doyle, M.P. and Padhye, V.V. (1989) *Escherichia coli*, in *Foodborne Bacterial Pathogens* (ed. M.P. Doyle). Marcel Dekker, Inc., New York, pp. 235–281.

Doyle, M.P. and Schoeni, J.L. (1984) Survival and growth characteristics of *Escherichia coli* associated with hemorrhagic colitis. *Applied and Environmental Microbiology*, **48**(4), 855–856.

Doyle, M.P. and Schoeni, J.L. (1987) Isolation of *Escherichia coli* O157:H7 from retail fresh meats and poultry. *Applied and Environmental Microbiology*, **53**(10), 2394–2396.

English, S. (1998) £2200 fine for *E. coli* outbreak that killed 20. *The Times*, London, 21 Jan 1998, p. 3.

Farmer J.J., III, Davis, B.R., Hickman-Brenner, F.W. *et al.* (1985) Biochemical identification of new species and biogroups of Enterobacteriaceae isolated from clinical specimens. *Journal of Clinical Microbiology*, **21**(1), 46–76.

Frank, J.F., Marth, E.H. and Olson, N.F. (1977) Survival of enteropathogenic and non-pathogenic *Escherichia coli* during the manufacture of Camembert cheese. *Journal of Food Protection*, **40**(12), 835–842.

Fukushima, H., Hashizume, T. and Kitani, T. (1997) The massive outbreak of Enterohemorrhagic *E. coli* O157 infections by foods poisoning among the elementary school children in Sakai, Japan in 1996. Abstract from *VTEC '97, 3rd International Symposium and Workshop on Shiga Toxin (Verocytotoxin)-producing* Escherichia coli *Infections*, June 22–26, 1997, Baltimore, Maryland, USA under the auspices of the Lois Joy Galler Foundation for Hemolytic Uremic Syndrome, Inc., USA.

Gammie, A.J., Mortimer, P.R., Hatch, L. *et al.* (1996) Outbreak of Vero cytotoxin-producing *Escherichia coli* O157 associated with cooked ham from a single source. *PHLS Microbiology Digest*, **13**(3), 142–145.

Gibbens, J. and Wray, C. (1997) Update on Vero cytotoxin-producing *E. coli* O157. *State Veterinary Journal*, **1** (April), 6–8.

Gibson, A.M. and Roberts, T.A. (1986) The effect of pH, water activity, sodium nitrite and storage temperature on the growth of enteropathogenic *Escherichia coli* and salmonellae in a laboratory medium. *International Journal of Food Microbiology*, **3**, 183–194.

Glass, K.A., Loeffelholz, J.M., Ford, J.P. *et al.* (1992) Fate of *Escherichia coli* O157:H7 as affected by pH or sodium chloride and in fermented, dry sausage. *Applied and Environmental Microbiology*, **58**(8), 2513–2516.

Griffin, P.M. (1995) *Escherichia coli* O157:H7 and other enterohemorrhagic *Escherichia coli*, in *Infections of the Gastrointestinal Tract* (eds M.J. Blaser, P.D. Smith, J.I. Ravdin, H.B. Greenberg and R.L. Guerrant). Raven Press Ltd., New York, pp. 739–761.

Gutierrez, E. (1996) Is the Japanese O157:H7 *E. coli* epidemic over? *The Lancet*, **348** (November 16), 1371.
Gutierrez, E. and Netley, G. (1996) Japanese *Escherichia coli* outbreak is still puzzling health officials. *The Lancet*, **348** (August 24), 540.
Hale, T.L., Echeverria, P. and Nataro, J.P. (1997) Enteroinvasive *Escherichia coli* in, Escherichia coli: *Mechanisms of Virulence* (ed. M. Sussman). Cambridge University Press, Cambridge, pp. 449–468.
Hancock, D.D., Besser, T.E., Rice, D.H. *et al.* (1997a) A longitudinal study of *Escherichia coli* O157 in fourteen cattle herds. *Epidemiology and Infection*, **118**, 193–195.
Hancock, D.D., Rice, D.H., Thomas, L.A. *et al.* (1997b) Epidemiology of *Escherichia coli* O157 in feedlot cattle. *Journal of Food Protection*, **60**(5), 462–465.
Hancock, D.D., Rice, D.H., Herriott, D.E. *et al.* (1997c) Effects of farm manure-handling practices on *Escherichia coli* O157 prevalence in cattle. *Journal of Food Protection*, **60**(4), 363–366.
Hara-Kudo, Y., Konuma, H., Iwaka, M., *et al.* (1997) Potential hazard of radish sprouts as a vehicle of *Escherichia coli* O157:H7. *Journal of Food Protection*, **60**, 9, 1125–1127.
Harrison, J.A. and Harrison, M.A. (1996) Fate of *Escherichia coli* O157:H7, *Listeria monocytogenes* and *Salmonella typhimurium* during preparation and storage of beef jerky. *Journal of Food Protection*, **59**(12), 1336–1338.
Harrison, J.A., Harrison, M.A. and Rose, R.A. (1997) Fate of *E. coli* O157:H7, *L. monocytogenes* and *Salmonella* spp. in reduced sodium beef jerky. *Journal of Food Protection Supplement B*, **60**, 20.
Health and Safety Executive (1996) *Interim Advice from the Health and Safety Executive on Laboratory Work with Verocytotoxin-producing* Escherichia coli. Health and Safety Executive, Sudbury, Suffolk.
Heuvelink, A.E., Wernars, K. and de Boer, E. (1996) Occurrence of *Escherichia coli* O157 and other Verocytotoxin-producing *E. coli* in retail raw meats in the Netherlands. *Journal of Food Protection*, **59**(12), 1267–1272.
Hinkens, J.C., Faith, N.G., Lorang, T.D. *et al.* (1996) Validation of pepperoni processes for control of *Escherichia coli* O157:H7. *Journal of Food Protection*, **59**(12), 1260–1266.
International Commission on Microbiological Specifications for Foods (1980a) *Microbial Ecology of Foods Volume 1: Factors Affecting Life and Death of Micro-organisms*. Academic Press London.
International Commission on Microbiological Specifications for Foods (1986) *Microorganisms in Foods 2. Sampling for Microbiological Analysis: Principles and Specific Applications,* 2nd edn. University of Toronto Press, Toronto.
International Commission on Microbiological Specifications for Foods (1996a) *Microorganisms in Foods 5. Microbiological Specifications of Food Pathogens*. Blackie Academic & Professional, London.
International Commission on Microbiological Specifications for Foods (1996b) *Establishment of Sampling Plans for Microbiological Safety Criteria for Foods in International Trade including Recommendations for Control of* Listeria monocytogenes, Salmonella enteritidis, Campylobacter *and Enterohaemorrhagic* E. coli. Codex Committee on Food Hygiene, 29th session, 21–25 October 1996, Agenda item 11, CX/FH 96/9 1-16. Codex Alimentarius Commission, Rome.

International Commission on Microbiological Specifications for Foods (1998) Micro-organisms in Food 6: *Microbial Ecology of Food Commodities*. Blackie Academic & Professional, London.

Jackson, T.C., Hardin, M.D. and Acuff, G.R. (1996) Heat resistance of *Escherichia coli* O157:H7 in a nutrient medium and in ground beef patties as influenced by storage and holding temperatures. *Journal of Food Protection*, **59**(3), 230–237.

Jaquette, C.B., Beuchat, L.R. and Mahon, B.E. (1996) Efficacy of chlorine and heat treatment in killing *Salmonella stanley* inoculated onto alfalfa seeds and growth and survival of the pathogen during sprouting and storage. *Applied and Environmental Microbiology*, **62**(6), 2212–2215.

Jones, D. (1988) Composition and properties of the family Enterobacteriaceae, in *Enterobacteriaceae in the Environment and as Pathogens, Proceedings of a symposium held at the University of Nottingham 7–9 July 1987* (eds B.M. Lund, M. Sussman, D. Jones and M.F. Stringer), The Society for Applied Bacteriology Symposium Series No. 17, *Journal of Applied Bacteriology Symposium Supplement*. Blackwell Scientific Publications, London, p. 8S.

Juneja, V.K., Snyder, O.P. and Marmer, B.S. (1997) Thermal destruction of *Escherichia coli* O157:H7 in beef and chicken: determination of *D* and *z* values. *International Journal of Food Microbiology*, **35**, 231–237.

Kauffmann, F. (1947) Review, the serology of the coli group. *Journal of Immunology*, **57**, 71–100.

Lehmann, K.B. (1893) *Methods of Practical Hygiene*, vol. 1. Kegan Paul, Trench, Trübner & Co. Ltd., London, p. 131.

Lin, C.-M., Fernando, S.Y. and Wei, Cheng-i. (1996) Occurrence of *Listeria monocytogenes*, *Salmonella* spp., *Escherichia coli* and *E. coli* O157:H7 in vegetable salads. *Food Control*, **7**(3), 135–140.

Line, J.E., Fain, A.R., Moran, A.B. *et al.* (1991) Lethality of heat to *Escherichia coli* O157:H7: *D* value and *z* value determinations in ground beef. *Journal of Food Protection*, **54** (October), 762–766.

Linton, A.H. and Hinton, M.H. (1988) Enterobacteriaceae associated with animals in health and disease, in *Enterobacteriaceae in the Environment and as Pathogens, Proceedings of a symposium held at the University of Nottingham 7–9 July 1987* (eds B.M. Lund, M. Sussman, D. Jones and M.F. Stringer), The Society for Applied Bacteriology Symposium Series No. 17, *Journal of Applied Bacteriology Symposium Supplement*. Blackwell Scientific Publications, London, pp.71S–77S.

Liu, M.N. and Berry, B.W. (1996) Variability in color, cooking times, and internal temperature of beef patties under controlled cooking conditions. *Journal of Food Protection*, **59**(9), 969–975.

MacDonald, K.L., Eidson, M., Strohmeyer, C. *et al.* (1985) A multistate outbreak of gastrointestinal illness caused by enterotoxigenic *Escherichia coli* in imported semi-soft cheese. *The Journal of Infectious Diseases*, **151**(4), 716–720.

Marier, R., Wells, J.G., Swanson, R.C. *et al.* (1973) An outbreak of enteropathogenic *Escherichia coli* foodborne disease traced to imported French cheese. *The Lancet*, December 15, 1376–1378.

Massa, S., Altieri, C., Quaranta, V. *et al.* (1997) Survival of *Escherichia coli* O157:H7 in yoghurt during preparation and storage at 4°C. *Letters in Applied Microbiology*, **24**, 347–350.

Maule, A. (1997a) Survival of the Verotoxigenic strain *E. coli* O157:H7 in laboratory-scale microcosms, in *Coliforms and* E. coli *– Problem or Solution* (eds D. Kay and C. Fricker). The Royal Society of Chemistry, London, pp. 61–65.

Maule, A. (1997b) Abstract from presentation: Survival of *Escherichia coli* O157 in laboratory-based model ecosystems and on surfaces, presented at the symposium *E. coli O157*, September 1997. Available from Campden and Chorleywood Food Research Association.

Mechie, S.C., Chapman, P.A. and Siddons, C.A. (1997) A fifteen month study of *Escherichia coli* O157:H7 in a dairy herd. *Epidemiology and Infection*, **118**, 17-25.

Mermin, J.H., Hilborn, E.D., Voetsch, A. *et al.* (1997) A multistate outbreak of *Escherichia coli* 0157: H7 infections asssociated with eating mesculin mix lettuce. Abstract from *VTEC '97, 3rd International Symposium and Workshop on Shiga Toxin (Verocytotoxin)-producing* Escherichia coli *Infections*, June 22-26, 1997, Baltimore, Maryland, USA under the auspices of the Lois Joy Galler Foundation for Hemolytic Uremic Syndrome, Inc., USA.

Miller, L.G. and Kaspar, C.W. (1994) *Escherichia coli* O157:H7 acid tolerance and survival in apple cider. *Journal of Food Protection*, **57**(6), 460-464.

Ministry of Agriculture, Fisheries and Food (1991) *Food Safety, A Guide from the Food Safety Directorate*. Ministry of Agriculture, Fisheries and Food, London.

Morgan, G.M., Newman, C., Palmer, S.R. *et al.* (1988) First recognised community outbreak of haemorrhagic colitis due to Verotoxin-producing *Escherichia coli* O157: H7 in the UK. *Epidemiology and Infection*, **101**, 83-91.

Morgan, D., Newman, C.P., Hutchinson, D.N. *et al.* (1993) Verotoxin producing *Escherichia coli* O157 infections associated with the consumption of yoghurt. *Epidemiology and Infection*, **111**, 181-187.

Mshar, P.A., Dembek, Z.F., Cartter, M.L. *et al.* (1997) Outbreaks of *Escherichia coli* O157:H7 infection and Cryptosporidiosis associated with drinking unpasteurized apple cider - Connecticut and New York, October 1996. *Journal of the American Medical Association*, **277**(10), 781-782.

Muir, R. and Ritchie, J. (1921) *Manual of Bacteriology*. Oxford University Press, London, pp. 353-359.

National Research Council (1985) An evaluation of the Role of Microbiological Criteria for Foods and Food Ingredients, US Subcommittee on Microbiological Criteria, Committee on Food Protection, Food and Nutrition Board, National Research Council, National Academy Press, Washington, DC.

Neaves, P., Deacon, J. and Bell, C. (1994) A survey of the incidence of *Escherichia coli* O157 in the UK dairy industry. *International Dairy Journal*, **4**, 679-696.

Nichols, G., Greenwood, M. and de Louvois, J. (1996) The microbiological quality of soft cheese. *PHLS Microbiology Digest*, **13**(2), 68-75.

Nutsch, A.L., Phebus, R.K., Riemann M.J. *et al.* (1997) Evaluation of a steam pasteurisation process in a commercial beef processing facility. *Journal of Food Protection*, **60**(5), 485-492.

Ørskov, F. (1984) Genus 1. *Escherichia* Castellani and Chalmers 1919, 941^{AL}, in *Bergey's Manual of Systematic Bacteriology*, vol. 1 (eds N.R. Krieg and J.G. Holt). Williams & Wilkins, Baltimore, pp. 420-423.

Orta-Ramirez, A., Price, J.F., Hsu, Y.-C. *et al.* (1997) Thermal inactivation of *Escherichia coli* O157:H7, *Salmonella senftenberg*, and enzymes with potential as time-temperature indicators in ground beef. *Journal of Food Protection*, **60**(5), 471-475.

Ostroff, S.M., Griffin, P.M., Tauxe, R.V. *et al.* (1990) A statewide outbreak of *Escherichia coli* O157:H7 infections in Washington State. *American Journal of Epidemiology*, **132**(2), 239-247.

Padhye, N.V. and Doyle, M.P. (1991) Rapid procedure for detecting enterohemorrhagic *Escherichia coli* O157:H7 in food. *Applied and Environmental Microbiology*, **57**(9), 2693–2698.

Palumbo, S.A., Pickard, A. and Call, J.E. (1997) Population changes and Verotoxin production of enterohemorrhagic *Escherichia coli* strains inoculated in milk and ground beef held at low temperatures. *Journal of Food Protection*, **60**(7), 746–750.

Park, H.S., Marth, E.H. and Olson, N.F. (1973) Fate of enteropathogenic strains of *Escherichia coli* during manufacture and ripening of Camembert cheese. *Journal of Milk and Food Technology*, **36**(11), 543–546.

Pennington, T.H. (1997) *The Pennington Group: Report on the Circumstances leading to the 1996 Outbreak of Infection with* E. coli *O157 in Central Scotland, the Implications for Food Safety and the Lessons to be Learned*. The Stationery Office Ltd, Edinburgh.

Phebus, R.K., Nutsch, A.L., Schafer, D.E. *et al.* (1997) Comparison of steam pasteurisation and other methods for reduction of pathogens on surfaces of freshly slaughtered beef. *Journal of Food Protection*, **60**(5), 476–484.

Reid, D. (1997) Introduction to the Second SCIEH Verocytotoxigenic *E. coli* Workshop, 31st January 1997. *Supplement to the Scottish Centre for Infection and Environmental Health Weekly Report*, **13**(1), 1.

Reitsma, C.J. and Henning, D.R. (1996) Survival of enterohemorrhagic *Escherichia coli* O157:H7 during the manufacture and curing of cheddar cheese. *Journal of Food Protection*, **59**, 460–464.

Report 71 (1994) *The Microbiology of Water. Part 1 - Drinking Water. Report on Public Health and Medical Subjects No. 71. Methods for the Examination of Waters and Associated Materials.* HMSO, London.

Riley, L.W., Remis, R.S., Helgerson, S.D. *et al.* (1983) Hemorrhagic colitis associated with a rare *Escherichia coli* serotype. *The New England Journal of Medicine*, **308**(12), 681–685.

Riordan, T., Gross, R.J., Rowe, B. *et al.* (1985) An outbreak of foodborne enterotoxigenic *Escherichia coli* diarrhoea in England. *Journal of Infection*, 11, 167–171.

Roberts, D., Hooper, W.L., and Greenwood, M.H. (eds) (1995) *Practical Food Microbiology. Methods for the Examination of Food for Micro-organisms of Public Health Significance. Section 6.6.* Public Health Laboratory Service, London.

Rowe, B. (1983) *Escherichia coli* diarrhoea. *Culture* **4**(1), 1–3 (published by Oxoid Ltd., Basingstoke, UK).

Ruiz, B.G.-V., Vargas, R.G. and Garcia-Villanova, R. (1987) Contamination on fresh vegetables during cultivation and marketing. *International Journal of Food Microbiology*, **4**, 285–291.

Salyers, A.A. and Whitt, D.D. (1994) *Bacterial Pathogenesis - A Molecular Approach*. ASM Press, Washington, DC. pp. 190–204.

Savage, W.G. (1912) *Milk and the Public Health*. Macmillan and Co. Ltd, London, pp. 237–240.

Semanchek, J.J. and Golden, D.A. (1996) Survival of *Escherichia coli* O157:H7 during fermentation of apple cider. *Journal of Food Protection*, **59**,(12), 1256–1259.

Sharp, J.C.M., Reilly, W.J., Coia, J.E. *et al.* (1995) *Escherichia coli* O157 infection in Scotland: an epidemiological overview, 1984-94. *PHLS Microbiology Digest*, **12**(3), 134–140.

Smith, H.W. (1961) The development of the bacterial flora of the faeces of animals and man: the changes that occur during ageing. *Journal of Applied Bacteriology*, **24**(3), 235–241.

Smith, H.R., Cheasty, T., Roberts, D. *et al.* (1991) Examination of retail chickens and sausages in Britain for Vero cytotoxin-producing *Escherichia coli*. *Applied and Environmental Microbiology*, **57**(7), 2091–2093.

Splittstoesser, D.F., McLellan, M.R and Churey, J.J. (1996) Heat resistance of *Escherichia coli* O157:H7 in apple juice. *Journal of Food Protection*, **59**(3), 226–229.

Steele, B.T., Murphy, N., Arbus, G.S. *et al.* (1982) An outbreak of hemolytic uremic syndrome associated with ingestion of fresh apple juice. *Journal of Paediatrics*, **101**, 963–965.

Stevenson, J. and Hanson, S. (1996) Outbreak of *Escherichia coli* O157 phage type 2 infection associated with eating precooked meats. *Communicable Disease Report*, **6**, Review No. 8, R116–R118.

Stewart, A.I., Jones, G.A., McMenamin, J. *et al.* (1997) Central Scotland *E. coli* O157 outbreak: clinical aspects (Monklands Hospital experience). *Scottish Centre for Infection and Environmental Health Weekly Report*, Number 1, No. 97/13, 9–11. Scottish Centre for Infection and Environmental Health, Glasgow.

Sumner, J. (1995) Victorian smallgoods industry. *Sunday Herald*, February 19, 45, Victoria, Australia.

Sussman, M. (1997) *Escherichia coli* and human disease, in Escherichia coli: *Mechanisms of Virulence* (ed. M. Sussman). Cambridge University Press, Cambridge, pp. 3–48.

Tilden, J., Young, W., McNamara, A.-M. *et al.* (1996) A new route of transmission for *Escherichia coli*: infection from dry fermented salami. *American Journal of Public Health*, **86**(8), 1142–1145.

Tkalcic, S., Harmon, B.G., Brown, C.A., *et al.* (1997) Effects of the rumen microenvironment on the growth and fecal shedding of *E. coli* 0157:H7. Abstract from *VTEC '97, 3rd International Symposium and Workshop on Shiga Toxin (Verocytotoxin)-producing* Escherichia coli *Infections*, June 22–26, 1997, Baltimore, Maryland, USA under the auspices of the Lois Joy Galler Foundation for Hemolytic Uremic Syndrome, Inc., USA.

Topley, W.W.C. and Wilson, G.S. (1929a) *The Principles of Bacteriology and Immunity*, vol. 1. Edward Arnold & Co., London, p. 446.

Topley, W.W.C. and Wilson, G.S. (1929b) *The Principles of Bacteriology and Immunity*, vol. 1. Edward Arnold & Co., London, pp. 1288–1293.

Tschäpe, H., Prager, R., Streckel, W. *et al.* (1995) Verotoxinogenic *Citrobacter freundii* associated with severe gastroenteritis and cases of haemolytic uraemic syndrome in a nursery school: green butter as the infection source. *Epidemiology and Infection*, **114**, 441–450.

Upton, P. and Coia, J.E. (1994) Outbreak of *Escherichia coli* O157 infection associated with pasteurised milk supply. *The Lancet*, **344** (October 8), 1015.

Viljanen, M.K., Peltola, T., Junnila, S.Y.T. *et al.* (1990) Outbreak of diarrhoea due to *Escherichia coli* O111:B4 in schoolchildren and adults: association of Vi antigen-like reactivity. *The Lancet*, **336**, 831.

Wallace, J.S., Cheasty, T. and Jones, K. (1997) Isolation of Vero cytotoxin-producing *Escherichia coli* O157 from wild birds. *Journal of Applied Microbiology*, **82**, 399–404.

Watanabe, H., Wada, A., Inagaki, Y. *et al.* (1996) Outbreaks of enterohaemorrhagic *Escherichia coli* O157:H7 infection by two different genotype strains in Japan, 1996. *The Lancet*, **348** (September 21), 831–832.

Weagant, S.D., Bryant, J.L. and Bark, D.H. (1994) Survival of *Escherichia coli* O157:H7 in mayonnaise and mayonnaise-based sauces at room and refrigerated temperatures. *Journal of Food Protection*, **57**(7), 629-631.

Wells, J.G., Davis, B.R., Wachsmuth, I.K. *et al.* (1983) Laboratory investigation of hemorrhagic colitis outbreaks associated with a rare *Escherichia coli* serotype. *Journal of Clinical Medicine*, **18**(3), 512-520.

Willshaw, G.A., Thirlwell, J., Jones, A.P. *et al.* (1994) Vero cytotoxin-producing *Escherichia coli* O157 in beefburgers linked to an outbreak of diarrhoea, haemorrhagic colitis and haemolytic uraemic syndrome in Britain. *Letters in Applied Microbiology*, **19**, 304-307.

Willshaw, G.A., Scotland, S.M. and Rowe, B. (1997) Vero cytotoxin-producing *Escherichia coli*, in Escherichia coli: *Mechanisms of Virulence* (ed. M. Sussman). Cambridge University Press, Cambridge, pp. 421-448.

Wilson, G.S. and Miles, A.A. (1964) *Topley and Wilson's Principles of Bacteriology and Immunity*, vol. 1, 5th edn. Edward Arnold, London, pp. 806-826.

Wray, C., McLaren, D.M. and Carroll, P.J. (1993) *Escherichia coli* isolated from farm animals in England and Wales between 1986 and 1991. *Veterinary Record*, **133**, 439-442.

Wright, D.J., Chapman, P.A. and Siddons, C.A. (1994) Immunomagnetic separation as a sensitive method for isolating *Escherichia coli* O157 from food samples. *Epidemiology and Infection*, **113**, 31-39.

Zhao, T. and Doyle, M.P. (1994) Fate of enterohemorrhagic *Escherichia coli* O157:H7 in commercial mayonnaise. *Journal of Food Protection*, **57**(9), 780-783.

Zhao, T., Doyle, M.P. and Besser, R.E. (1993) Fate of enterohemorrhagic *Escherichia coli* O157:H7 in apple cider with and without preservatives. *Applied and Environmental Microbiology*, **59**(8), 2526-2530.

Zhao, T., Doyle, M.P., Shere, J. *et al.* (1995) Prevalence of enterohemorrhagic *Escherichia coli* O157:H7 in a survey of dairy herds. *Applied and Environmental Microbiology*, **61**(4), 1290-1293.

Index